Scienceworks!
A Guide to
Science Instruction..
... with a difference!
Using Content
as a Base for the
Growth of...
Science Skills
Science Concepts
Science Attitudes

SCIENCE WORKS!

ISBN-13: 978-0615666648
ISBN-10: 0615666647

PRINTING HISTORY
First Printing: JUNE, 1987
Second Printing: APRIL, 1988
Revised Edition: El Amarna Publishing: MAY, 2013

SCRIPT: SUSAN SMILY
ARTWORK: LOUAYNE RHODE
BRAINPOWER: THE RAT (SUE) and THE FROG (LOUAYNE)

SPECIAL NOTE: A fair amount of what is in this document is based on curriculum material that I received while on teacher exchange in Western Australia. I was extremely impressed with both the level of their commitment to Elementary Science, and the support material that they were developing. The underlying philosophy was the same as that which has been emerging across North America, and the material, while simple and succinct, covered many of the areas of Elementary Science that need to be addressed in any effective program. I therefore borrowed liberally from their guide books, while trying to adapt, add to, and restructure the material in order to develop a light-hearted yet thorough examination of the fundamentals of a sound program. I hope that I have succeeded. In addition, I have found that these fundamentals are what is needed today, as we try to educate new students for the 21st century. Sound teaching principles never go out of style. (Susan Smily, 2013)

Dedication
For my Mum -
who introduced me to Dickens and Darwin.

CAST OF CHARACTERS

Follow us all through the pages as we look at Science Teaching
from the perspective of 2 students—Sue (The Rat) and Louayne (The Frog)!

Jergen Jargon will look at BIGGGG words...
Our teachers, Mr. and Mrs. Beamish will offer timely advice...
The snake will try to work his way up to a level of understanding...
Guffaw Punster will offer some terrible jokes...
The Footnote just hangs around looking important!

Scienceworks
Hi, Sue What's THAT you've got?
THE WORKS
SCIENCEWORKS ?!? What are THEY?
My teacher keeps our science classes in Gear with them!
bad pun
Do you mean...
... in THAT box...
I bet my teacher would LOVE them!
I can show you how they WORK!
CONTENT
Can We look at this one?
THE WORKS
Yes, but just one at a time. We mustn't overload the children...

TABLE OF CONTENTS

SECTION 1 - GETTING READY

INQUIRY LEARNING

INVESTIGATION

USING THE LIBRARY

LESSON PLANNING

GROUP WORK

EVALUATION STRATEGIES

WHAT'S IT ALL ABOUT, ALFIE?

SCIENCEWORKS
IS FORMED FROM MANY INDIVIDUAL COMPONENTS,
EACH ONE A SUBSECTION
OF ONE OF FOUR BASIC THEMES.

CONTENT Concepts SKILLS attitudes

They may appear to be unique,
and capable of being managed as separate entities.

BUT...
who knows the limitless ways in which they may be combined?

We are faced with insurmountable opportunities

This publication is divided into **six** thematic sections
but only in order to provide you with an easy reference point.
If we select only one at a time for examination in our instruction,
we severely narrow our view
and limit the potential of both our program and our students.

We have examined over 50 separate components of science instruction,
planning, and implementation.

First, we set you up with an overview of some of the basics:
Inquiry Learning, *Investigation*, *Research*, *Lesson Planning*, *Group Work*, and *Evaluation*.

Then, we look at the whole range of the *Affective Domain*,
and how a *Scientific Attitude* ties in with other positive attitudes.

We look at *Basic Science Concepts*,
and then show you a few techniques for building *Schemata* –
taking the step beyond *Concept Development*.

We examine a variety of *Thinking Skills*,
and ways to get our brains in gear.

We try to show how all the *Process Skills* are part of the whole.

We take a brief look at **eight** different *Content* areas,
giving consideration to the scope of each area
and offering some basic teaching strategies.

In the *Activity* mini-units
we have used a mix and match technique
that allows us to examine all the components with the students
over the course of the school year,
while not overloading them with too much at one time.

(We hope that we've created a resource tool, and not a monster!)

INQUIRY LEARNING - WE DEVELOP EXPERIENCES

If we are given the opportunity to test concepts, we learn to make sense of the information that we get ***and*** we learn how to work with that information.

(*Footnote says, "Ignore Side A. Take a nap. Most students would!)

The test might be:

Make a quiz board that lights up when you test the right answer with the wire.
or
Make a model office building with electricity within.

That means that if we have the opportunity to test our concepts as they are learned, we can develop the schemata necessary to work at higher levels of thinking.

INVESTIGATION - WE LOOK FOR ANSWERS

PROVIDING INVESTIGATIVE OPPORTUNITIES

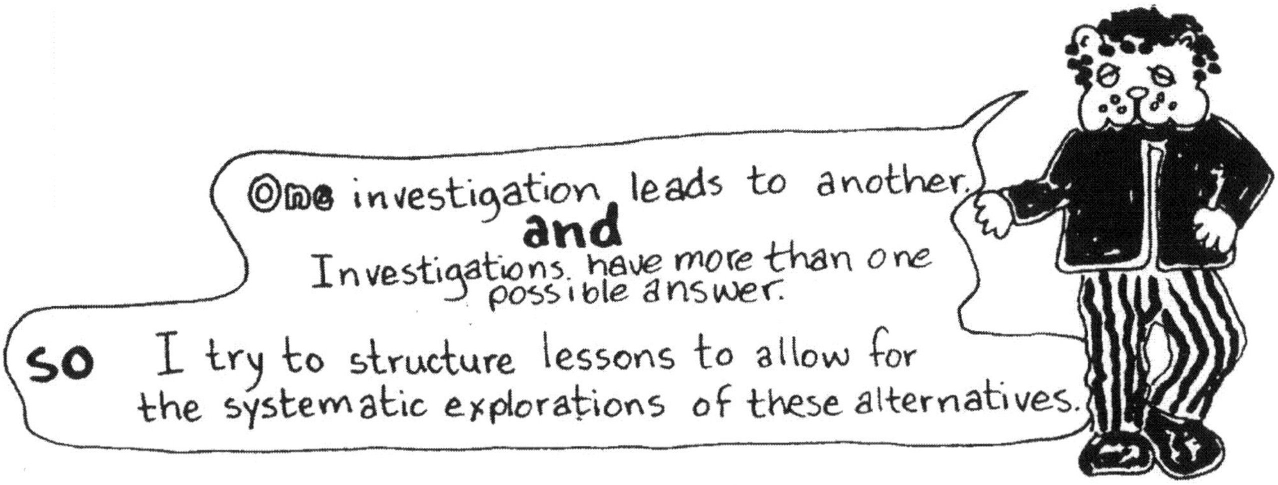

INQUIRY LEARNING INVOLVES HANDS-ON EXPERIENCES AND INVESTIGATIONS.

THE VALUE OF INVESTIGATION

Investigation gives us
an opportunity
to find answers for ourselves.

We like to explore things
that puzzle us,
until we gain an understanding
that satisfies us.

SET YOUR OUTCOMES CONSIDERING OUR ABILITIES!

If we can explore,
we learn faster and better.

And so our teacher
provides challenges for us
and the opportunity to discover,
test, evaluate, and research
information.

USING THE LIBRARY - WE BROADEN OUR RESOURCE BASE

THE LIBRARY IN OUR SCHOOL IS A RESOURCE CENTER...

When I need a lot of information,
I can find a lot of information on the internet.
But I make sure the articles I read are well documented,
and I learns the rules for using what I find,
so I don't violate the copyright laws.

I also like looking at encyclopedias,
and many other sources that are just as useful and a lot more interesting!

The magazine rack is especially useful in Science.
Ranger Rick, National Geographic World, Owl, Chickadee—
all of these provide up-to-date and visually stimulating material on all sorts of topics.

Our library also has a terrific vertical file—
there's a picture for just about everything.

Our school district has lots of DVDs and CD ROMs
on many science topics.
And our school library has a few, too.

And of course there are books!

Lots and lots of fantastic books!

USING THE LIBRARIAN – WE UTILIZE EXPERTISE

Don't forget the best resource that any library has.
That's the teacher/librarian, of course!
The librarian can help me—and my teacher—
to find out about a lot of different things.

The teacher/librarian in my school likes to help out
through cooperative planning activities.

How?

There are various ways that this can be done.

➥ Two teachers in the classroom—
helping individuals and sharing the load during experimental sessions.

➥ Splitting the class and having instruction in both the class and library.
We may use ability grouping or just a simple two-way split.

Getting Ready

Our librarian needs to know what is happening,
what the unit objectives are,
and what our final product will look like,
in order to be fully involved in the whole unit.

The librarian's not a mind-reader, though,
and I can't always explain exactly what my teacher was asking for.
I sometimes sort of know what I'm doing, but might not be able to explain it.
If the librarian and my teacher have discussed it in advance,
then they both know what's up.

So, we all get together right at the start of a unit
and discuss what's going to be happening.

That way, even if the librarian is not directly involved in the teaching of the unit,
we can get help when we show up at the library.

One other thing—our teacher checks with the librarian
about convenient times BEFORE sending us to the library.
You see, our librarian really hates to have to turn us away
or to be seen as unavailable.

Library Skills - We Learn Strategies

I've learned a lot of library skills
so that I can find what I want, when I want.

I've learned how to use our library's cataloging system.

I know how to find different subject areas—
I look for the signs!

I've learned how to use an index.
I know where to find the reference books.

I've learned how to take notes,
how to use the computer and other AV material,
and how to write a bibliography.

AND...
I KNOW MY WAY TO THE NEAREST PUBLIC LIBRARY.

THE BEST REPORT EVER!

Research is very important, since I can't always examine everything first-hand!
But when I do research, I don't just copy the article out of the encyclopedia or off the internet.
I'm trying to do my "BEST REPORT EVER",
and I have some guidelines that can help me to do that.

Seven Steps to Good Research

STEP 1: I think carefully about my topic, and then I choose four or five questions to which I'd like the answers.

For example:
1. How do astronauts get to the International Space Station?
2. What kind of food do they eat?
3. How do they send messages back to Earth?
4. Why doesn't the Soyuz burn up when it's landing?

STEP 2: Now I collect information the same way that a reporter does.
I go to reference books, I ask questions, I look at pictures.
I take notes, but I don't copy straight from the books or the internet, because those aren't my words, and the teacher and I both know that!

STEP 3: Next I make a rough draft, and I try to write the information down in such a way that I could explain what I have learned to someone 1-2 grades younger than I am.
I choose my words very carefully, and I check to make sure that it makes sense to me. That way it can make sense to other people, too.

STEP 4: I read my draft aloud to a friend and find out if it's interesting and if it makes sense.
I also listen to myself as I read it, so that I know where to put commas and periods.

STEP 5: With a red pen, I proofread for spelling and punctuation.
After I've done my own proofreading, I take my work to my teacher and ask for a little bit of help in assuring that everything is all right.

STEP 6: I decide how I want to present my project.
Would it look best as a booklet, poster, chart, or something else?
I consider the layout, borders, corner designs, headings, and drawings.
I want it to look attractive and be easy to read.

STEP 7: Now I can rewrite my project from my edited copy, hand it in, and wait for the compliments!

LESSON PLANNING - TO MAKE IT WORK

THERE ARE MANY THINGS TO CONSIDER IN PLANNING.

Science lessons don't always follow their intended path. Our investigations sometimes lead us in a direction which the teacher can't foresee, but which is still considered valuable.

This does not mean, however, that thoughtful planning is not necessary. It enables the teacher to provide us with materials and guidance to follow those individual paths.

Potentially sound lessons with interesting activities can flounder, or can have minimal impact because of incomplete planning.

LONG RANGE – THE WHOLE UNIT

What investigations will be an essential part of the unit?

What investigations will be optional?

What anticipated directions might further investigations develop?

How can those investigations be related to other areas of the curriculum?

What particular skills, attitudes, or concepts is this unit designed to develop?

How will my teacher assess which of us have adequately developed these objectives?

Teacher's Note: The skills sections in this book—5A and 5B— have general and summative evaluation criteria.

Short Range – Today's Lesson

What tasks will we perform in this lesson?

Will all of us be investigating the same problems, or will there be a range of problems?

How will the tasks be presented to us?

Will we be working as a total class unit, in small groups, or individually?

If groups are to be used, what size will they be, and how will they be formed?

What materials will be needed?

How will these be obtained?

How will they be distributed to us?

Will we be expected to record our findings? If so, why and how?

Will we be able to cope with the tasks? The form of presentation? The organization?

Teacher's Note: All of the Units at the back of this book have a "Unit Objectives" page.

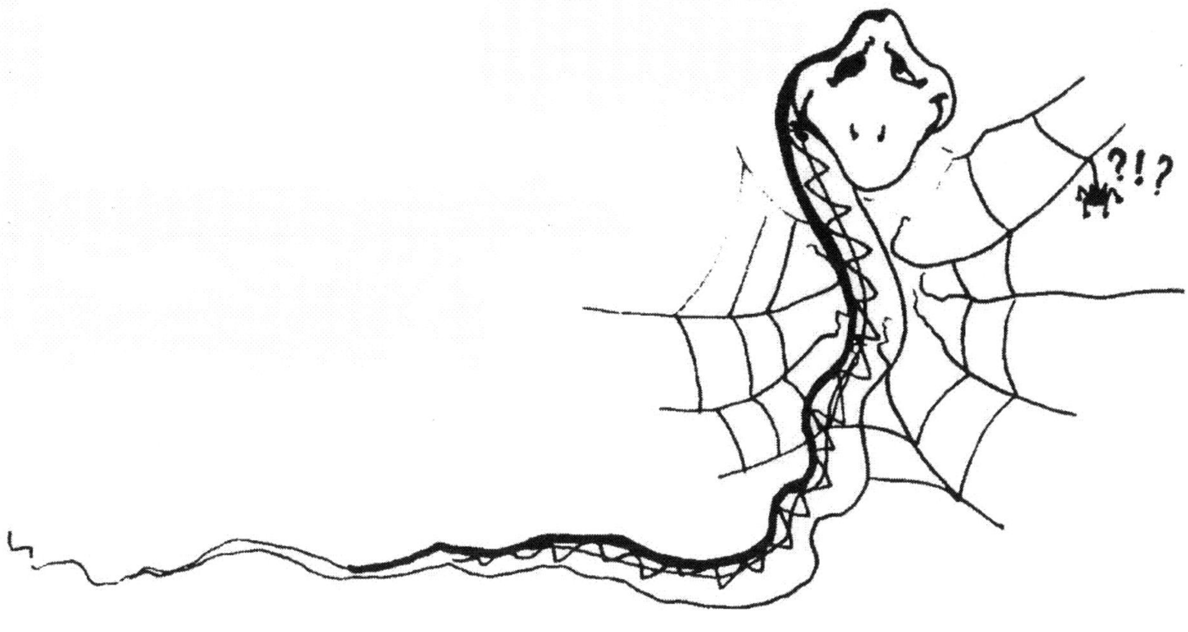

GROUP WORK - TO SHARE THE LOAD

GROUPING

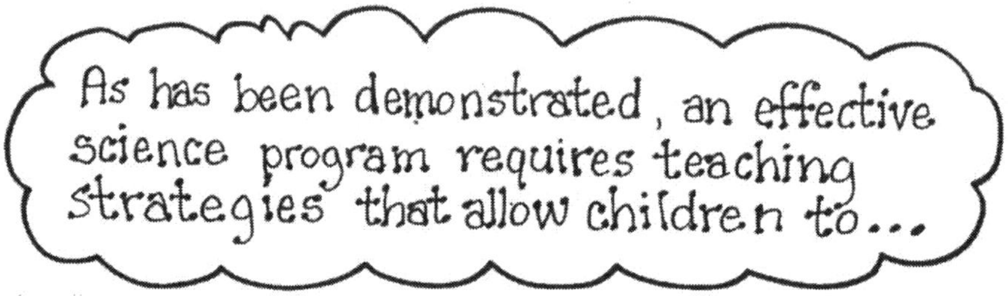

- handle material,
- work and talk with other children,
- use initiative in seeking answers to questions.

Many teachers and children are unfamiliar with this type of teaching/learning situation. We'd like to share with you some strategies that we have found useful in grouping.

YOU SEE, we have seen many teachers make a fundamental error.

They place children into a group situation with a pile of materials and an open-ended task when the children have no prior experience with this type of learning. They are then expected to cope effectively and quietly.

CHAOS is the predictable result.

Therefore, it makes sense to plan a sequence of experiences that enables both the teacher and the children to become used to working in this way.

My teacher starts by discussing the differences between PLAY and SCIENCE LEARNING ACTIVITIES.

We discover that in order to learn from our activities, we need to pay close attention to what we are doing and to what is happening.

DISRUPTIVE CHILDREN

When disruptive children interfere with the group, there are several ways of relieving the problem.

My teacher might:

Physically space the children so that noise and movement are less obvious.

Allow some children to work outside the classroom, in the library, in the cloakroom, etc.

Group disruptive children with children with whom they will cooperate.

Allow disruptive children to work on a "special" task, which is different from the activities being done by other students.

If possible, obtain the support of a colleague—possibly the librarian, or a parent volunteer, or a mentor from a higher grade—who can assist until the remainder of the children are working effectively, so you will have more time to cope with the ones that need special monitoring or assistance.

MAINSTREAMED CHILDREN

Similar procedures will help you cope with mainstreamed children.

All children are allowed some success: those who cannot do investigations effectively might be able to be trained as equipment monitors.

Given the opportunity, many children will show leadership in helping others. My teacher will often discuss the situation and allow us to suggest alternatives.

Developing Investigative Skills

As we become more experienced with doing science activities, our teacher gives more and more of the responsibility to us.

LEVEL 1: WE ALL WORK ON THE SAME TASK.

The task is divided by the teacher into a series of simpler activities requiring about 5 minutes each.

Each activity is presented to the whole class, with fairly precise directions given.

We are able to talk to a partner, but we all have our own materials.

When completed, the results of the task are discussed before the next activity is presented.

LEVEL 2: WE ALL WORK ON THE SAME TASK.

The task is divided into 15 minute activities.

Some choice is available to us as to how the task is performed.

We are able to talk to a partner, but we all have our own materials from a shared pool.

LEVEL 3: WE ALL WORK ON THE SAME TASK.

The task is black-boarded, together with three alternative approaches.

We work with a partner, with shared materials and a joint investigation.

DON'T RUSH THE PROCESS - GIVE EACH STEP SUFFICIENT TIME TO BE EFFECTIVE.

LEVEL 4: WE WORK ON ONE OF 3-5 DIFFERENT TASKS.

All the tasks have a common theme
and are presented on activity cards.

We choose how to conduct the investigation
and decide what materials we will need.

We work in groups of three, with shared materials and a joint investigation.

LEVEL 5: WE SELECT ONE OF MANY DIFFERENT ACTIVITIES.

All the tasks are presented on activity cards.

We choose the task we would like to work on, plan how to conduct the investigation,
and decide what materials we will need.

We work in groups of three, with shared materials and a joint investigation.

LEVEL 6: WE IDENTIFY INVESTIGATIONS OF OUR OWN.

We decide how the investigation will be conducted.
We select, locate, or construct our own materials.

We form our own groups of a manageable size.

The task is subdivided by each group into component investigations.

EVALUATION STRATEGIES - TO MAKE IT COUNT

WHY EVALUATE?

MY TEACHER HAS GOOD REASONS FOR EVALUATING ME...

I can be provided with immediate feedback,
and thus have my successes reinforced
or get help in understanding my failures.

Any of my classmates who need special help
can be easily identified –
as well as the areas in which they need
remediation or enrichment.

Our parents can find out how we're really doing –
nobody likes guesswork
on interview night!

The teacher learns a lot
about our different strengths and weaknesses –
and assesses both teaching effectiveness
and how much we got out of the lesson.

Who Do We Evaluate For?

WE ALL WANT TO KNOW SOMETHING...

THE PARENT

What is my child doing?

Is my child coping with the work?

Is my child happy, confident, satisfied?

How can I help my child to improve?

THE CHILD

Am I performing well in Science?

What are my strengths?

Where do I need to improve?

What does my teacher want me to learn or to do?

THE TEACHER

Are the children improving?

In what aspects is each child improving?

What children need special assistance?

Am I catering for individual needs?

Are my objectives being achieved? How can I tell?

Are my objectives reasonable, appropriate, worthwhile?

How Do We Evaluate?

MY TEACHER EXAMINES A WIDE VARIETY OF EVALUATION TECHNIQUES.

The evaluation technique depends on:

- what is being evaluated
- who the evaluation is for
- individual, school and district priorities
- our developmental stage

Tests

Tests are easy to devise and easy to mark and give scores that parents and children can understand.

ACTIVITY (PRACTICAL) TESTS:

We use given material to demonstrate knowledge and understanding of properties, functions, etc.

PICTURE TESTS:

We complete drawings to demonstrate our knowledge.

PEN AND PAPER TESTS:

We write answers to questions to demonstrate our knowledge.

HOWEVER...

My teacher realizes that it is very difficult to devise a test that will assess anything other than knowledge.

Observer Records and Checklists

A scale is devised to differentiate levels of achievement.

Records such as this are quite subjective,
but if the criteria are well defined in advance,
teachers can avoid glib report card comments.

Remember that as teachers, they are,
by training and expertise,
experts in subjective assessment,
and many of the most important objectives
of education are evaluated subjectively.

Diary / Anecdotal Records

Sometimes there are some observations about me
or another student which are worth making a note or two about.

This does not imply that an essay needs to be written about each of us for every activity.

Final Product Assessment

Our teacher keeps in mind the criteria that are being assessed.
The assessment looks beyond neatness, presentation, and (apparent) amount of work.

I get to try an alphabet soup of different techniques.

AUDIO FILES
✸
BLOGS
✸
BOOKS - INTEREST/EXPERIENCE
✸
CHART, COLLAGE, COLLECTION
✸
DEMONSTRATION, DIARY, DIORAMA, DISPLAY, DRAMATIZING, DRAWING
✸
FILM
✸
GRAPHS: BAR, PICTURE, LINE
✸
INTERVIEWS
✸
MAPS, MOBILES, MODELS (STILL/WORKING)
✸

NEWSPAPER REPORTS
✸
PAMPHLETS, PASTED PICTURES, PHOTOGRAPHS, POEMS, POLYSTYRENE CUTOUTS, POSTERS
✸
QUESTIONNAIRES
✸
REPORTS (ORAL/WRITTEN)
✸
SLIDES, SCALE DRAWINGS, SONGS, STORIES, SURVEYS
✸
TV SCROLLS, TAPES, TIME LINES
✸
VIDEOS, DVDs
✸
WEBSITES

A B C D E F G H

Why Do We Use Different Techniques?

It's more fun for all of us if many different means of evaluation are used.

No single form of assessment
is best for all situations.

My teacher knows that a variety of methods,
used on a variety of occasions, will give a broader
and more useful collection of information.

From this, patterns will emerge
which indicate my development.

It doesn't need to make heavy time demands;
most of the recording is done during the lesson.

**MY TEACHER KNOWS WHAT'S BEING DONE,
WHAT'S BEING PLANNED,
AND THE REASONS FOR DOING IT!**

There is a commitment to a set of science goals.

The development of appropriate
SKILLS, ATTITUDES, and CONCEPTS is planned,
with careful consideration of the objectives
as well as of our needs and interests.

We know what the objectives are.

Evaluation is undertaken directly
in accordance with the objectives.

More than factual recall is evaluated.

The CONTENT chosen is designed
to develop the process SKILLS, foster the ATTITUDES,
and strengthen the CONCEPTUAL understandings
proposed within the planning objectives.

SECTION 2 - DEVELOPING A SCIENTIFIC ATTITUDE

A SCIENTIFIC ATTITUDE

CURIOSITY

CARING

PERSISTENCE

INVENTIVENESS

CRITICAL THINKING

THE AFFECTIVE DOMAIN

GENERAL ATTITUDE DEVELOPMENT

ATTITUDES - TO GROW WITH

CURIOSITY – WE QUESTION

CURIOSITY
IS A NATURAL WONDER
ABOUT THE WORLD AROUND US.

Developing Curiosity

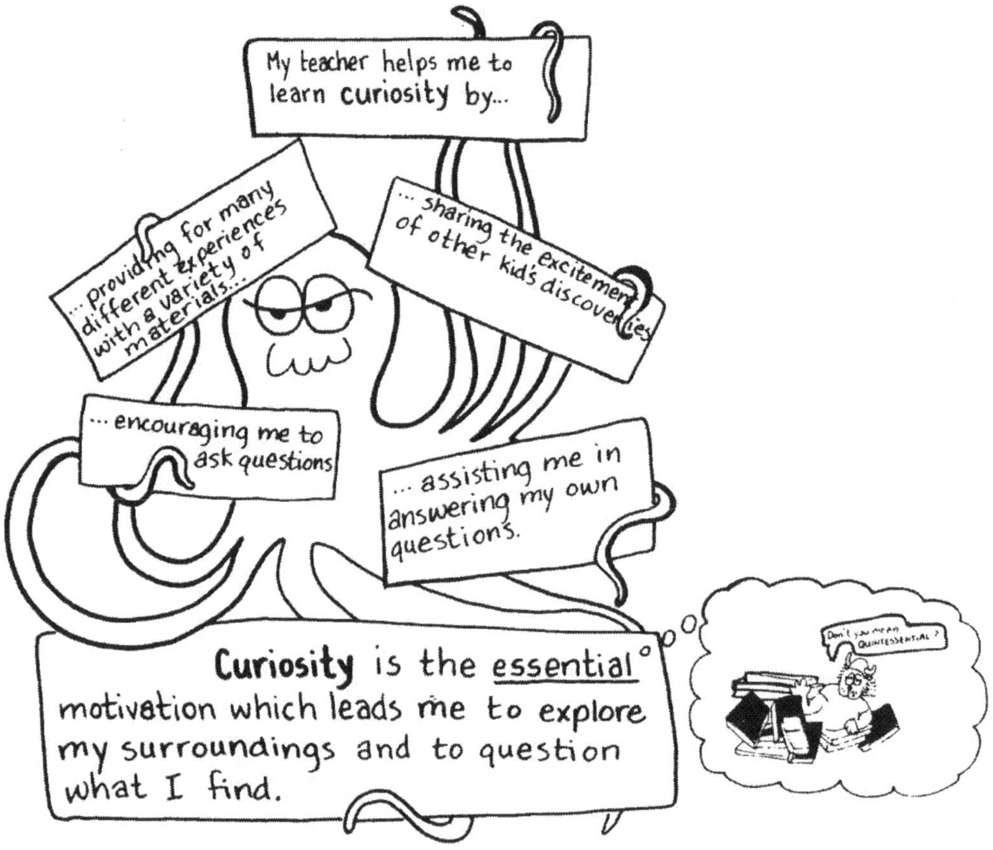

Demonstrating Curiosity

I SHOW MY SCIENTIFIC CURIOSITY BY...

- enjoying science investigation.
- showing an interest in new or unusual things.
- asking questions on a variety of topics.
- showing excitement about my observations.
- obtaining new ideas and new insights from things I know, or things that are new.
- preferring to find things out for myself.
- exploring things until I feel I understand them.
- trying to discover and explain causes and relationships.

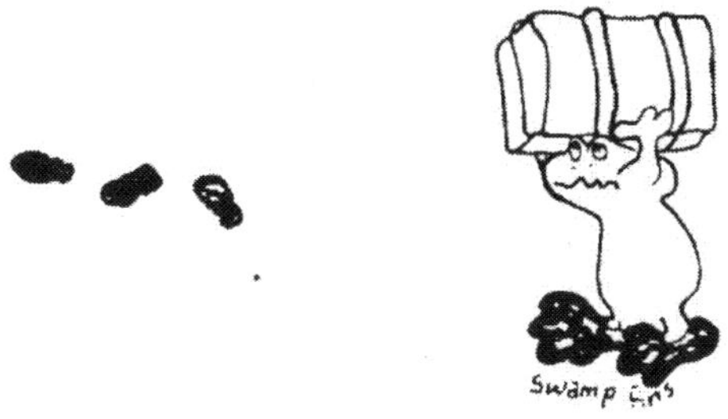

Evaluating Curiosity

LEVEL 1: At a SUPERIOR level, I exhibit these characteristics constantly; I am inquisitive about all investigations; I explore beyond the requirements of the assignment; I reach conclusions that will help me to investigate further.

LEVEL 2: On the AVERAGE, I am interested, but not quite as inquisitive; I follow the progress of the experiment, but I do not extend the results into another activity.

LEVEL 3: My performance is SATISFACTORY if I am interested during the activity but am generally content to know the facts as given to me during instruction.

LEVEL 4: My attitude would be UNSATISFACTORY if I showed little interest in new things and seldom questioned any event.

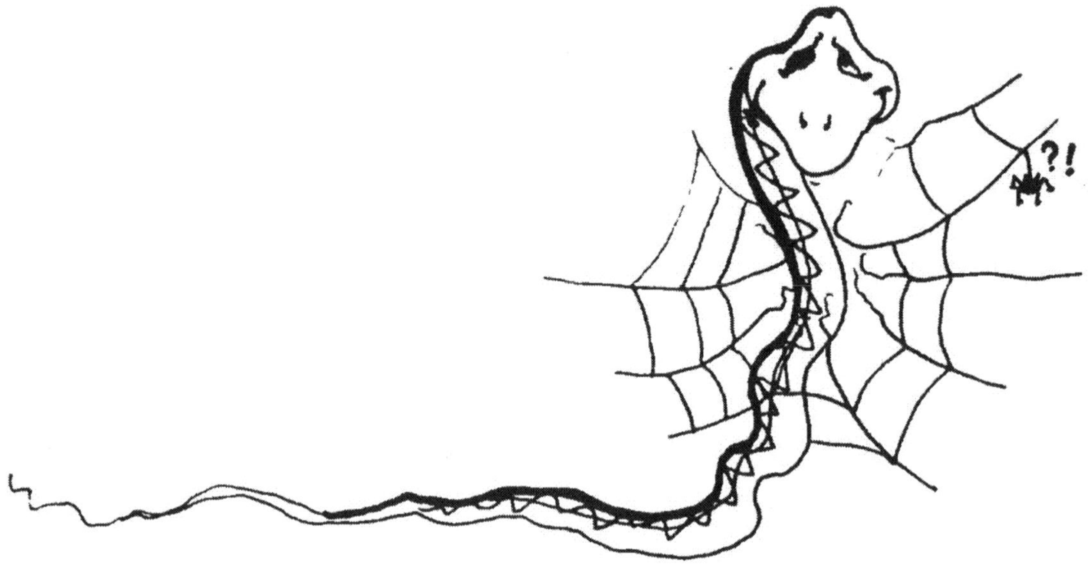

CARING - WE SHOW CONCERN

CARING
IS A CONCERN FOR LIVING THINGS AND THE QUALITY OF THE ENVIRONMENT.

Developing Caring

MY TEACHER HELPS ME DEVELOP A CARING ATTITUDE BY...

- providing me with opportunities to care for animals and plants.
- setting up situations in which I can observe all the things that happen (the repercussions) when the environment is misused.

CARING ALSO INVOLVES THE DEVELOPMENT OF AN AWARENESS OF SELF AND SURROUNDINGS.

MY TEACHER ALSO HELPS ME TO...

- examine the effects of my behavior on my environment and on others.
- examine the effects of other's activities on the environment and on each other.
- examine the effects of the earth's processes on the environment and on human activity.

Demonstrating Caring

I SHOW A CARING ATTITUDE BY...

- caring for myself, my friends, my environment.
- accepting the responsibility of caring for the natural and man-made environment.
- suggesting ways of maintaining and improving my surroundings.
- understanding the interaction of human behavior, natural phenomena and environmental concerns.
- understanding the concept of coexistence for all.
- growing as an earth-conscious citizen.

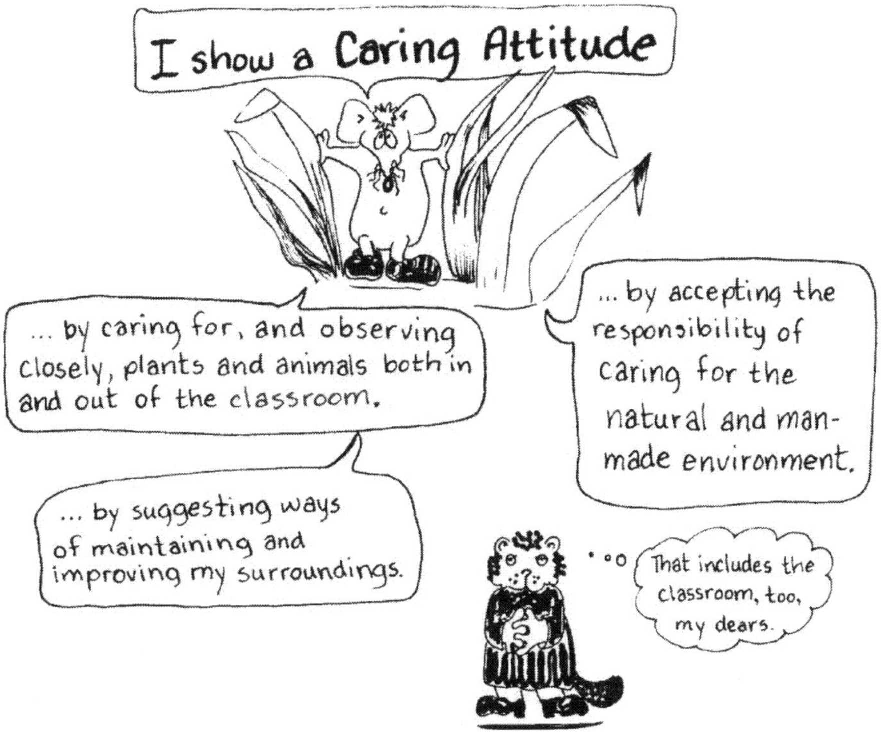

Evaluating Caring

LEVEL 1: At a **SUPERIOR** level, I am fully aware of my environment and recognize my responsibilities; I am always responsible in my treatment of animals; I fully accept the concept of "the whole earth"; I work toward becoming a world citizen.

LEVEL 2: On the **AVERAGE**, I accept responsibility for my immediate environment and occasionally initiate improvements; I generally handle animals with care; I am aware that human behavior has an effect on my surroundings and that natural processes affect the way humans live; I accept that people all over the world have a responsibility toward the proper use of the earth.

LEVEL 3: My performance is **SATISFACTORY** if I participate in classroom initiated environmental activities; I am generally careful with pets and other animals; I am able to explain some ways in which my behavior affects the environment, and I am willing to take some responsibility for caring for my immediate surroundings.

LEVEL 4: My work would also be **UNSATISFACTORY** if I am generally self-centered; I am unaware of the effect of my behavior on the environment; I show little concern for others; I leave responsibility to others; I am unconcerned with the well-being of plants or animals.

Persistence – We Don't Give Up

PERSISTENCE
IS THE WILLINGNESS TO PURSUE A PROBLEM UNTIL A PERSONALLY SATISFYING SOLUTION IS FOUND.

Developing Persistence

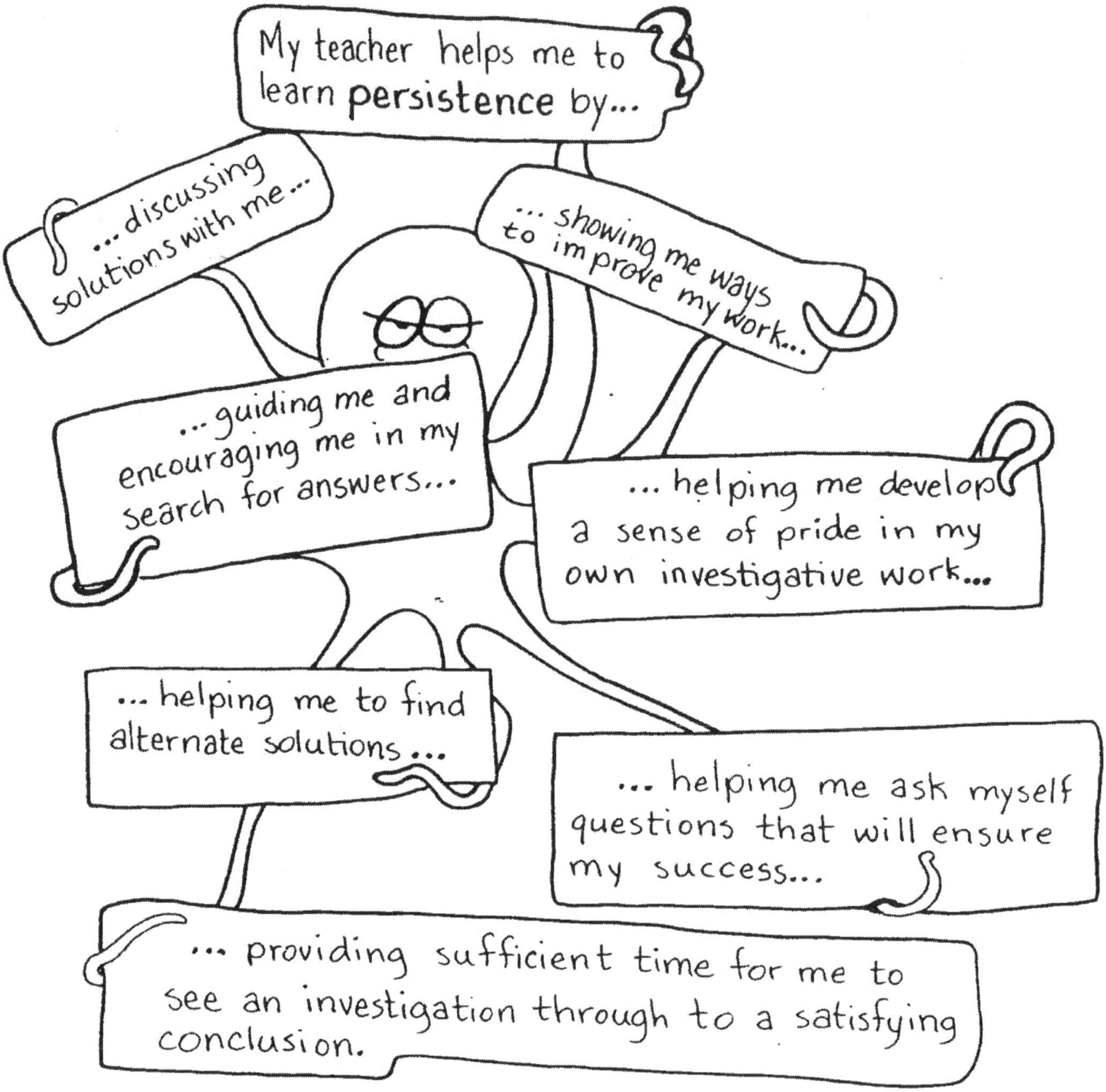

Demonstrating Persistence

I SHOW PERSISTENCE BY...

- staying with a problem or an experiment until I am completely satisfied with my results.

  This may mean that I have to redo an activity if the results seem unsatisfactory or contradictory.

  It does not mean that I have to keep on doing an experiment until I get the "right" answer, however, since there is never really a "right" answer!

- attempting to overcome any difficulties that may arise by myself, before I ask for help.
- being prepared to risk failure in trying an approach that I consider worth trying.
- changing my approach if I'm unsuccessful but not giving up.

Evaluating Persistence

LEVEL 1: At a **SUPERIOR** level, I exhibit these characteristics constantly; I am willing to stay with a task until it is completed; I am not satisfied with less than a complete answer.

LEVEL 2: On the **AVERAGE**, I work to the completion of a project, but may need occasional reminder; I do not let minor set-backs impede my progress.

LEVEL 3: My work is **SATISFACTORY** if I work at projects during the time assigned but generally do not carry projects beyond the main requirements; I am able to work past most frustrations.

LEVEL 4: My work would be **UNSATISFACTORY** if I am unable to stay with a problem or project for even a short period of time; I give up easily when frustrated.

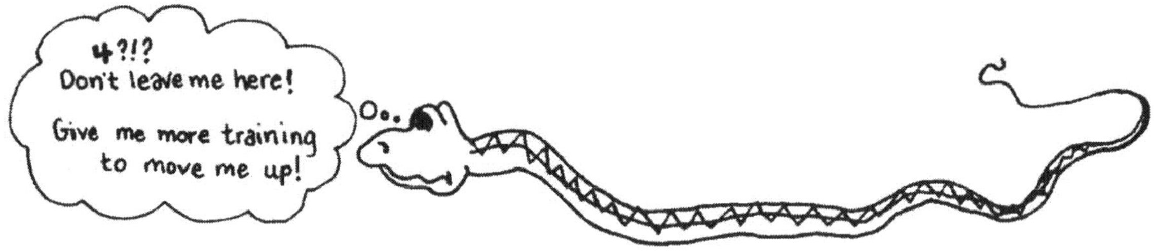

INVENTIVENESS - WE CREATE

INVENTIVENESS
IS THE WILLINGNESS AND ABILITY
TO CONSIDER SEVERAL
ALTERNATIVE APPROACHES TO A PROBLEM.

Developing Inventiveness

MY TEACHER HELPS ME TO DEVELOP A HIGHER LEVEL OF INVENTIVENESS BY...

- posing problems that require me to think and plan, rather than simply to follow directions.
- building my creative thinking skills.
- encouraging me to propose alternative ways of solving a problem.
- allowing me to approach a problem with a method that differs from the teacher's.

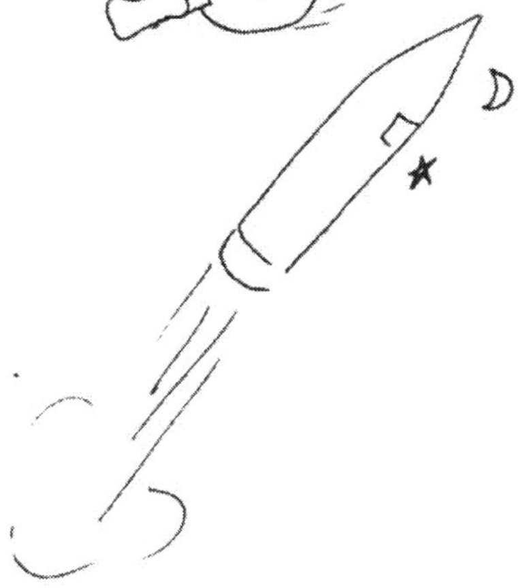

Demonstrating Inventiveness

I SHOW INVENTIVENESS BY...

➤ suggesting alternate ways of solving a problem or extending an activity.

➤ drawing upon, but not limiting myself by, past experience and relevant knowledge.

➤ obtaining or improving equipment required for an activity.

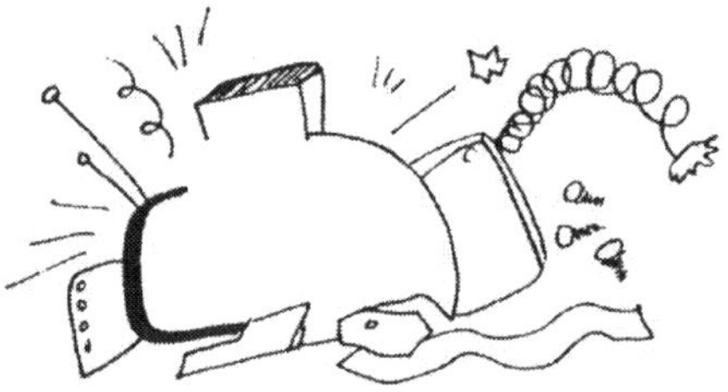

➤ not requiring constant teacher direction.

Evaluating Inventiveness

LEVEL 1: At a **SUPERIOR** level, I show well-developed creative-thinking skills; I offer many alternatives during class discussions; I am inventive and creative in my approach to all aspects of school work; I consider many options in the completion of science activities; I extend my investigations beyond the assignment.

LEVEL 2: On the **AVERAGE**, I participate in classroom planning, contributing ideas and alternatives; I am able to solve most of the problems associated with completing independent activities; I consider some follow-ups to my class work.

LEVEL 3: My work is **SATISFACTORY** if I try to add ideas to classroom discussions; I try one or two different solutions when faced with a problem; I am able to collect most of the equipment necessary for my activity.

LEVEL 4: My work would be **UNSATISFACTORY** if I do not try to contribute to planning; I do not consider any extension to activities; I leave it to others to gather equipment; I stop work when confronted by a problem.

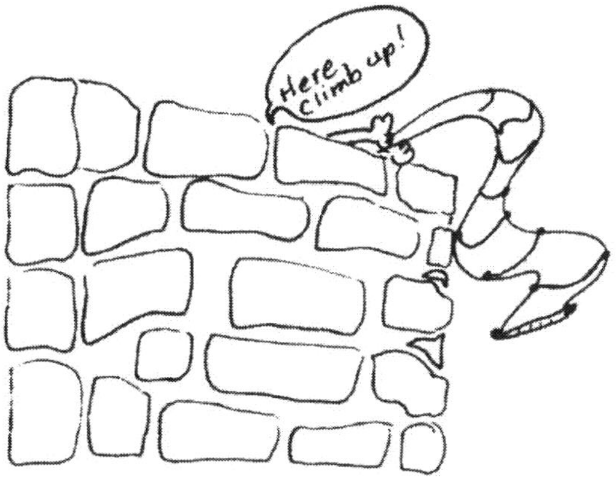

CRITICAL THINKING – WE PONDER

CRITICAL THINKING
IS AN ATTITUDE OF OPEN-MINDEDNESS TO IDEAS, BUT WITH A SKEPTICISM TOWARD SOLUTIONS NOT SUPPORTED BY EVIDENCE.

Developing Critical Thinking

MY TEACHER HELPS ME LEARN CRITICAL THINKING BY...

- encouraging me to question the findings of my classmates and those of the teacher.
- pointing out situations where conflict occurs in evidence and in giving explanations about observations.
- accepting unexpected answers, seeking explanations for my reasoning and encouraging further investigation into an event or observation rather than rejecting any erroneous findings as wrong out of hand.
- encouraging me to defend my views against criticism if I am confident that the view is based on sound reasoning.

Demonstrating Critical Thinking

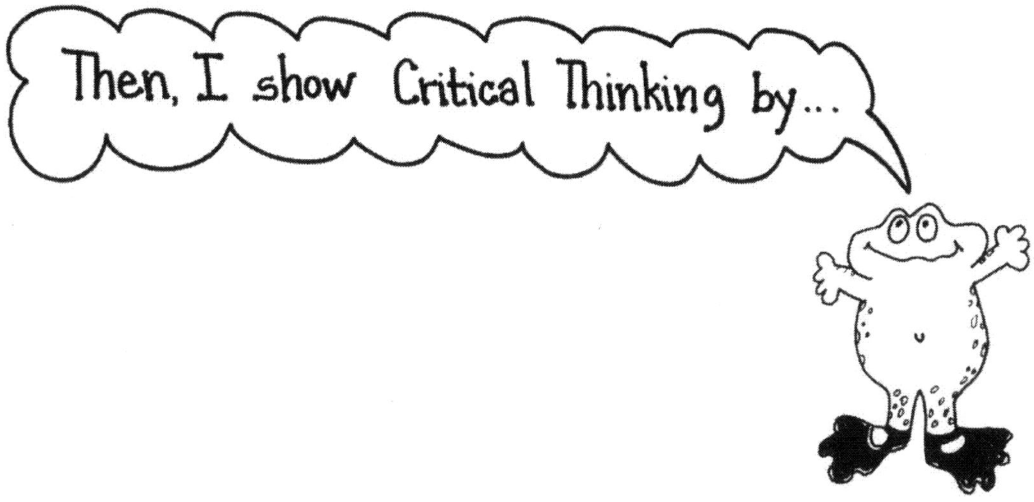

➤ questioning my own and others findings.

➤ identifying errors in my own approach and learning from them.

➤ trying to obtain data to prove or disprove guesses and explanations.

➤ forming conclusions which are consistent with evidence and challenging conclusions not based on evidence.

➤ being prepared to hold an unpopular view.

➤ distinguishing between an observation and an inference and between a conclusion based on evidence and one based on feeling.

Evaluating Critical Thinking

LEVEL 1: At a **SUPERIOR** level, I am able to analyze effectively all aspects of an investigation, distinguishing between guesses and factual explanations, between observations and inferences, between emotive and logical reasoning; I am willing to challenge conclusions that are not based on evidence.

LEVEL 2: On the **AVERAGE**, I can distinguish between an inference and an observation; I can form some conclusions based on my observations; I can find data to support my explanations; I can counter challenges to my conclusions.

LEVEL 3: My work is **SATISFACTORY** if I can separate most inferences from observations; if I understand that conclusions should be based on evidence.

LEVEL 4: My work would be **UNSATISFACTORY** if I guess at everything; I cannot support by conclusions; I don't bother to get substantiating data for my conclusions; I react emotionally to challenges to my observations.

ATTITUDES AND THE AFFECTIVE DOMAIN - UNDERSTANDING EMOTIONS

THE AFFECTIVE DOMAIN

Having a Scientific Attitude goes far beyond merely "enjoying science class". It involves the development of a clarity of thought, a view toward the subtle, the intricate, the definite; and at the same time toward the obvious, the simple, the divergent.

Having a Scientific Attitude means I am able to approach all aspects of my life, all the problems I may encounter, with an open and inquiring mind. I am more likely to be persistent in my endeavors, more likely to explore new avenues, more likely to be aware of the effects of my activities on others.

With a scientific approach toward activities and problem solving, I recognize the need for developing my process skills and my thinking skills, and I work toward their full and extensive utilization.

Science activities give me and my classmates the opportunity to develop and clarify our own attitudes and values relating to ourselves and our environment.

We are able to recognize that others may have attitudes and values that are different from, but equally as defensible as our own, and we take into account their ideas and attitudes when we express, defend, and modify our own ideas and attitudes.

Affective learning goes on all the time—even if our teacher is not planning for it, since we get a feeling for everything we do, as we do it.

Affective learning, however, can be inferred from my behaviors, and my teacher can adjust any instructional techniques to take advantage of positive growth or to forestall negative growth.

Therefore, attitudes toward a subject must be considered, since my early feelings and opinions about it have a long-term effect on my later learning in the subject.

As I develop long-term affective learnings, I acquire:

Appreciations	*Behaviors*	*Preferences*
Emotions	*Opinions*	*Interests*
Habits	*Feelings*	*Beliefs*
Tastes		*Values*

GENERAL ATTITUDE DEVELOPMENT

In addition to the specific attitudes noted on the previous pages,
there are several general attitudes that my teacher can foster within the Science Program.

Honesty

Rationality

Objectivity

Suspended Judgement

Open-Mindedness

Humility

MY TEACHER TRIES TO REMEMBER THAT EMOTIONS ARE AT LEAST AS IMPORTANT IN A LEARNING SITUATION AS INTELLECT.

Honesty

I am reluctant to compromise with the truth.
I consciously report all observations in a truthful manner.

My teacher encourages all of us to report our observations honestly, even if they seem to be incorrect. We are given open-ended experiments, in which we don't know the answer for a particular set-up, so that I won't feel there is only one possible answer.

This gives me more confidence and a greater sense of independence.

Rationality

The search for plausible solutions should not be influenced by superstitious explanations.

As I develop rationality, I seek natural causes for events, and I am cautious about permitting decisions to be affected by personal likes, fears, anger, or ignorance.

My teacher cultivates this attitude through an examination of the rationality of certain common superstitions and by setting up simple experiments to verify or disprove them.

I am presented with situations in which logical reasoning proves superior to superstitious explanations.

I become more aware of the difference between fact and opinion.

I develop a realization that facts must be backed up by proof.

Objectivity

As I develop a sense of objectivity,
I am not guided by personal feelings and do not let my feelings interfere
with the impersonal judgement needed
when I am collecting and interpreting data.

My teacher arranges situations where the inferences that are made
may reflect variations due to individual differences and personal feeling,
so that we can look at both sides of the issue
and seek objective solutions.

My teacher stresses the importance of unbiased research.

Suspended Judgement

My teacher helps me develop a reluctance
to forming a generalization based upon inadequate evidence.

My hypotheses are tentative,
and my knowledge base is constantly being revised.

The teacher, therefore, develops situations in which we are given choices –
we can make judgements based on varying amounts of data.

We learn that insufficient data leads to inaccurate results,
whereas suspended judgement, yielding more data, leads to success.

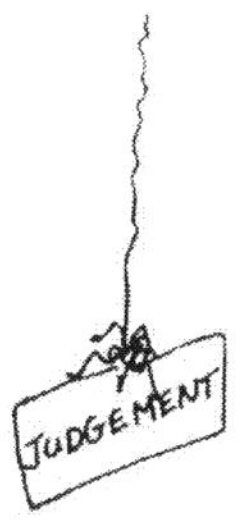

Open-Mindedness

I exhibit a willingness to change my mind in light of new evidence.

My teacher provides experiences in which
I have the opportunity to consider novel hypotheses and unique procedures.

I am encouraged to seek unorthodox solutions to traditional problems.

My teacher discusses some historical examples with us,
in order to illustrate how a well-accepted hypothesis
has changed because of new evidence.

Humility

As we mature, we develop a recognition of our own limitations
as well as the limitations inherent in science itself.

Limitations of the human senses
as well as limitations offered by the instruments of science
can be understood first-hand through direct experience
and vicariously through discussions and reading.
Our teacher provides both.

THE SCIENTIFIC MIND

Rationality	*A scientific person is rational in the approach to learning,*
Curiosity	*shows an eager desire to learn,*
Open-Mindedness	*has a mind receptive to new ideas,*
Suspended Judgment	*is unprejudiced,*
Objectivity	*is undistorted by emotion or personal bias,*
Critical Thinking	*makes careful, analytical evaluations,*
Persistence	*is unrelenting in efforts to learn,*
Inventiveness	*is a resourceful,*
Caring	*concerned individual,*
Honesty	*who is fair, sincere,*
Humility	*and unpretentious.*

SECTION 3 -
EXAMINING BASIC SCIENCE CONCEPTS

GENERAL UNDERSTANDINGS

INTELLECTUAL DEVELOPMENT AND GROWTH

DEVELOPING SCIENCE CONCEPTS

INDIVIDUAL DIFFERENCES

CONCEPTS ABOUT ANIMALS

CONCEPTS ABOUT PLANTS

CONCEPTS ABOUT HEALTH

CONCEPTS ABOUT MATTER

CONCEPTS ABOUT ENERGY

CONCEPTS ABOUT THE EARTH

CONCEPTS ABOUT SPACE

CONCEPTS ABOUT TECHNOLOGY

SCHEMATA

INTEGRATION

CONCEPTS - TO INTERNALIZE OUR LEARNING

CONCEPTS ARE GENERAL UNDERSTANDINGS

My teacher knows that Conceptual Development
is a vital component of Elementary Science.

My classmates and I will find it difficult to remember the multitude of facts
that we are going to be confronted with over the course of 12 years of schooling.

But if all of these facts are associated with a few fundamental concepts,
then an understanding will grow that has a much greater long-term significance
for us than the mere acquisition of knowledge.

**A CONCEPT IS DEVELOPED
FROM A RANGE OF EXPERIENCES—
NOT LEARNED.**

Therefore, the more experiences that I have,
the more refined my understanding of any concept will be.

As the information comes in, it is useless to my brain unless it is presented in a way
that I can connect it to previous information and develop my schemata from it.

My future learning must expand from and enhance
the schemata that I already possess.

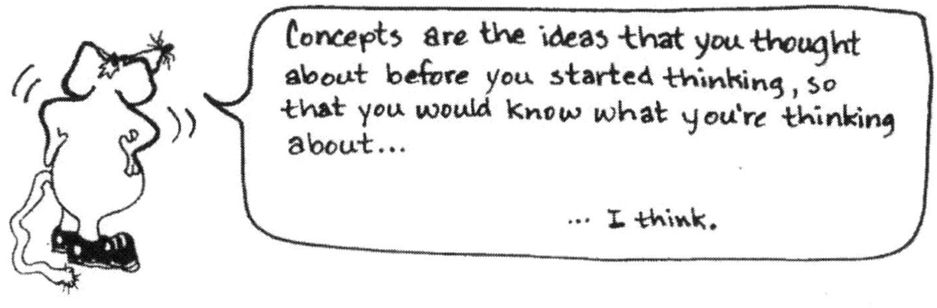

Concepts Are Connections

We are constantly making connections,
and it is through this process that we learn,
and that our learnings become fully integrated.

This is not abstraction, it is based on first-hand experience.
The closer you bring learnings to my immediate environment,
the more sense they make, and the more likely I am to integrate them.

I need to take part—passive observation will NOT be enough.

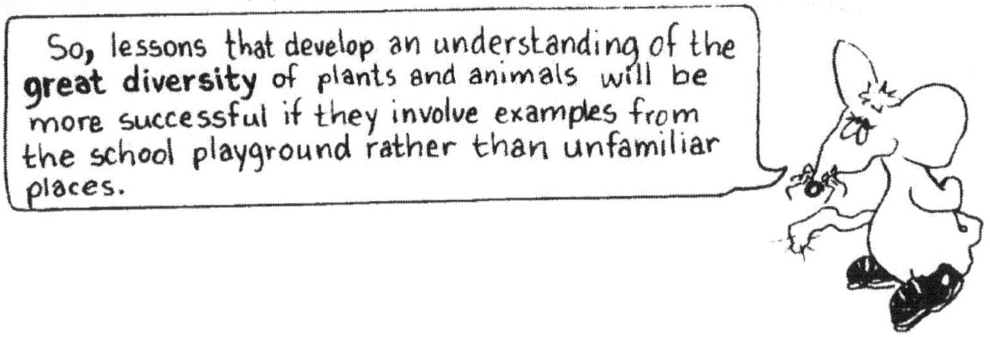

Concepts Are Different From Facts

While I will gain factual knowledge about many different topics,
that knowledge is really incidental to the development of general understandings.

It is my store of conceptual understandings that will stay with me
long after the facts are forgotten.

Conceptual Growth Is Sequential

In elementary school I do little if any abstract thinking;
it takes first-hand experience and manipulation of my environment
for me to be able to understand the concepts that my teacher may be teaching me.

So, while I can understand that solids melt and liquids evaporate,
I am not yet ready to learn about the theory of molecular movement.

My teacher knows that it is important to understand
my level of conceptual awareness
about a particular topic before beginning instruction.

Intellectual Development and Growth

Jean Piaget's theory of intellectual development in children can help my teacher
to understand my level of conceptual understanding.

Each stage of learning is part of my development
and necessary to the development of the next stage.
Before I proceed from one level to the next I need to integrate and utilize the one before.
My teacher tries to be aware of where I am, as well as where I am going.

Setting the Stage

The Sensori-Motor Stage
(Age: Birth - 2 Years)

At first, an object exists only if I can see it or feel it.
Hidden objects are forgotten, or I might randomly search for them.

There is no other time but the present, and no other space than that which I can see.

I must do something in order to imagine doing it;
but I am learning language, how to control my muscles, basic socialization skills;
and I am constantly using my senses to develop the basic knowledge and understandings
that form the basis and foundation of the next stage of development.

The Pre-Operational Stage
(Age: 2-7 Years)

At this stage, I am very egocentric, and I do not use logical operations in my thinking.
I see the world as it affects me—subjectively.

I am perceptually oriented—the way things look to me is the way they are,
and I make judgements based on my observations.

I operate intuitively and depend upon trial and error to make corrections.

I can describe variables (properties or aspects of a phenomenon or object),
but I concentrate on one variable at a time
and usually the most visually dominant one.

I am not yet very good at realizing that an object has several properties,
and so it is difficult for me to combine parts into a whole.

I am fairly good at arranging objects into simple series,
but longer series confuse me,
and if you give me a new object,
I may not be able to put it into its proper place in the series.

Space is no larger than my own community;
time is measured in days, hours, and seasons.

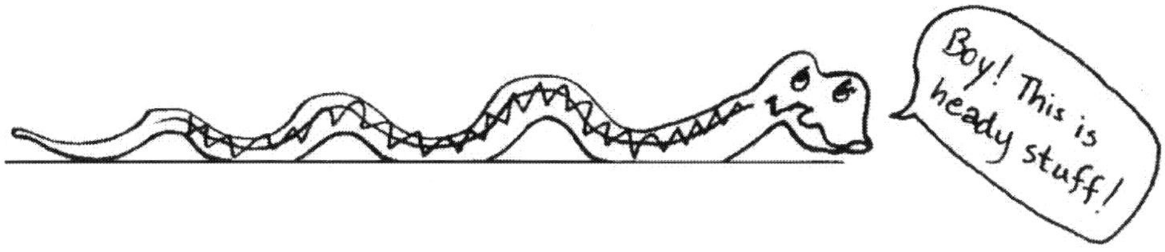

The Concrete Operational Stage
(Age: 7-11 Years)

I can now perform logical operations,
but my thinking is still concrete rather than abstract.

I can solve physical problems;
I can make objects fit together properly by moving them around;
I can hold two or more variables in my mind;
I can understand multiple relationships and classifications;
I can combine parts into a whole.

I can start to understand geographical space and historic time.

I develop the concept of reversibility
and can reverse processes when the size or shape of matter is changed.

The Formal Operational Stage
(Age: 11-15 Years)

My thinking has shifted from the concrete to the more formal and abstract.

I can now relate one abstraction to another,
and I am able more and more to think conceptually.

I have started to develop hypotheses,
and I can test them with controlled experiments.

I can reflect upon the solution to a problem
and the thought process that I used to solve it.

For example, when I was in Grade 1,
a rock was something that I fell down on,
and it hurt my knee,
or it was something that broke a window.

By Grade 4, I was more aware of the fact
that rocks were a part of the earth
and that there were many different kinds of rocks.

As I became older I gained more knowledge about earth processes,
which allowed me to place rocks into a much wider context
in my understanding of the earth.

Developing Science Concepts

My teacher can ascertain my level of understanding of a topic
merely by asking me a few questions at the beginning of each unit.

My teacher can then combine this with an understanding of individual differences
to establish which stage I am in
and can give help so that I can learn within that framework.

In Science,
I need to be helped to pass progressively from one stage to the next,
through the provision of challenging problems
that encourage me to move into the next stage of growth.

Learning by discovery helps me to "learn how to learn".

However, I cannot learn the variety of problem-solving techniques
and inquiry methods without experiencing them first-hand.

Hands-on activities are more rewarding than text book study or demonstrations,
are more easily remembered and recalled,
and allow me to apply my learnings to new situations,
thus acquiring new concepts.

Individual Differences

Each student has something to offer.

Science can provide a wide range of learning activities
that meet the varying needs of the students.

Individual learning is easily integrated in a science program
as we perform our activities and make observations and inferences.

Gifted Children

It is important that my teacher recognize the many differences in ability, interest, and need that my classmates have.

Some of our group are gifted—they are quick to learn, and they can remember concepts more easily and with sharper detail.

They have a longer attention span, so they are able to concentrate much more effectively and show a high degree of persistence.

They are very curious and original, they are critical of their own work, and they take the initiative in problem solving.

They can deduce the scientific concepts that underlie their observations and the facts that they are given and can derive other concepts from the deduced ones.

Special Needs Students

Some students, on the other hand, require much more time to grasp a concept
and may forget or become vague about what they have learned.

They may have a much shorter attention span
and give up when they encounter difficulties.

My teacher gives them constant guidance and encouragement,
knowing that often they would rather carry out directions
than plan their own activities.

They need a lot of concrete, firsthand experience, and considerable repetition,
as they are generally unable to arrive at key concepts on the basis of observation.

They are not able to evaluate their work critically
and thus have difficulty in correcting their mistakes.

My classmates in these categories may well be
at completely different developmental stages from the rest of us
and may require different instructional and evaluation strategies.

EXAMINING THE CONCEPTS - TO UNDERSTAND THE WORLD

LOOKING AT THE BASIC CONCEPTS

Most concepts that explain events and processes in science
can be grouped into one of seven broad areas of conceptual understanding.

Change

The environment, and everything in it, is in a constant state of change.

Interaction

Things are constantly interacting with each other, usually causing change.

Cause and Effect

Changes in substances or organisms can usually be traced to a particular cause.

Variety / Commonality

All things differ, yet they can be grouped
in terms of some similar characteristics or properties.

Interdependence

The survival of any living thing is interwoven
with other living things and the physical surroundings.

Continuity

Life and the forces at work in the environment do not cease.

Conservation

Substances can change but do not vanish.

Looking at Concepts

Some Concepts About Animals

➢ Animals grow and change.

➢ The rate of growth differs for different animals.

➢ The activities of animals may cause changes in their habitats.

➢ Animals respond to seasonal changes.

➢ Different creatures show considerable variety in the different stages of their life-cycles.

➢ Many animals construct shelter: nests, lodges, etc.

➢ Some animals compete for, and establish, territory.

➢ Some animals can live in a wide variety of environments, some need a specialized environment.

➢ The life processes, or activities that living things do in order to live, both cause, and are caused by, constant change in the organism.

➢ Our natural environment involves a complex interrelationship between plants, animals, soil, and weather.

Some Concepts About Plants

➢ Plants are living things that grow and change.

➢ There are many kinds of plants, which have a variety of physical characteristics.

➢ All plants, and plant parts, have basic characteristics, but these characteristics differ from plant to plant.

➢ Some plants, or parts of plants, are harmful to humans.

➢ Plants have needs—water, sunlight, food, etc.

➢ Plants and parts of plants change as they grow.

➢ Plants respond to stimuli, such as light, water, and gravity.

➢ Water is absorbed through the roots, toward the leaves.

➢ The fruit of a plant contains seeds.

➢ Vegetables are living plants.

➢ Some plants are a source of food for some animals. Ultimately, all animals are dependent on plants for food.

➢ Plants reproduce in a variety of ways.

➢ Parts of a plant can be used to propagate new plants.

➢ When plants are propagated, the new plant has the same characteristics as the old plant.

➢ Plants transpire from their leaves. Plant transpiration is an important part of the water cycle.

➢ Plants have adaptations to control transpiration, thus aiding survival in arid areas.

➢ Plants are important to the environment in all areas, including those which may be inhospitable.

Some Concepts About Health

➢ People are living things that grow and change.

➢ There are many kinds of people,
who have a variety of physical characteristics.

➢ Since people have physical and behavioral adaptations
suited to their environments,
they can live in a wide variety of environments.

➢ People construct shelters and make changes to their environment,
which assist them to survive.

➢ Some people compete for and establish territories,
but most cooperate in temporary or permanent associations.

➢ Some people depend on other people for food and protection.

➢ People and other animals are in many ways interdependent.

➢ Overpopulation, over-consumption,
and pollution may endanger communities.

Some Concepts About Matter

➢ Objects can float or sink in water.

➢ A pencil marks a paper.

➢ Water can erode the land.

➢ Substances are classified as liquid, solid, or gas.

➢ Some substances can be easily changed from one state to another.

➢ Substances change in different ways.
When heated, substances may turn to liquid, burn,
change color, or not change at all.

➢ Matter can change from solid to liquid or to gas
by being heated or cooled.

➢ Physical changes take place in matter
when the state is changed by heating or cooling,
or when a change is made in the shape or size of the object.
Physical changes are, to some extent, reversible.

➢ Chemical changes cause the creation of a new substance,
with properties that are different from the old substance.

➢ Chemical changes are evident in burnt wood,
soured milk, and digested food.

➢ Energy (heat, light, electricity, etc.)
is needed for or produced by chemical changes.
Chemical changes are not reversible.

Some Concepts About Energy

➢ The sun provides light and heat for all living things.

➢ The gain or loss of heat in a substance can cause changes in the nature of the substance. (e.g., color, volume, plant and animal behavior, etc.)

➢ Burning requires fuel, oxygen, and the appropriate ignition temperature.

➢ Objects, living things, and substances can be made to move, respond, or change in other ways by the use of energy.

➢ Light and heat come from many sources; the most important is the sun.

➢ Sound can be produced by both living and non-living objects.

➢ Temperature is the measurement of the heat of a substance.

➢ Some substances allow a flow of electricity; others do not.

➢ Light, heat, sound, and electricity can be produced in a variety of ways.

➢ Energy can be produced in a particular form by changing it from some other form: e.g., wind into mechanical energy.

➢ When coal is burned, the chemical energy in the fuel is released and changed into heat energy.

➢ The heat energy can be used to turn water into steam. The steam can turn a turbine to produce mechanical energy.

➢ The turbine can run an electric generator that changes mechanical energy into electrical energy.

➢ Electrical energy may be changed into heat energy, light energy, or sound energy.

Some Concepts About the Earth

➢ The earth was formed in the same way as the sun,
the planets, and other heavenly bodies.

➢ The earth was formed from the same materials as the sun.

➢ The crust is a very thin outer layer of the earth.

➢ The inside of the earth is very hot.

➢ The solid part of the earth's crust
is made up of great masses of hard material called rock.

➢ Most of the rock on the surface is covered with soil.

➢ Rocks are not made by and do not come from, living things.

➢ Rocks are made up of minerals.

➢ The solid parts of the earth can move,
and this movement can change the appearance of the surface of the earth.

➢ Forces on and within the earth change the shape of the earth,
wearing the surface away and building it up again.

➢ Soil is made up of tiny grains of rock and mineral.

➢ Soil becomes fertile when humus—the remains of dead animals
or plants—is added to it.

➢ Air masses move across the earth, changing the weather
➢ Weather changes with the seasons.

➢ The water cycle allows water to be constantly reused,
filling lakes and replenishing the oceans
and providing rain and moisture for plant and animal life.

Some Concepts About Space

- The planets orbit the sun because of the gravitational pull of the sun; the rotation of the earth on its axis causes day and night.
- Energy from the sun allows life to exist on earth.
- The sun heating the earth causes weather and climates.
- The earth's revolution around the sun causes the year.
- The spin of the earth causes day and night.
- The tilt of the earth to the sun causes seasons.
- The pull of gravity from the moon on the earth's oceans causes tides.
- Gravity and inertia keeps planets in orbit around the sun and their moons in orbit around them.
- Eclipses are caused when the sun casts the earth's shadow on the moon or the moon's shadow on the earth.
- Sunspots are caused by magnetic storms in the sun.
- Comets tails are melted gases that have been vaporized.

Some Concepts About Technology

- Technology has both good and bad effects on society.
- Society controls the development of technology.
- Society influences and responds to scientific development.
- Technology is both a cause and a result of scientific activity.
- Technology changes the skills and knowledge needed to do many things in life.

SCHEMATA – TO PULL IT ALL TOGETHER...

SCHEMATA
ARE INTEGRATED WAYS
OF LOOKING AT THE WHOLE ENVIRONMENT–
THE WHOLE WORLD.

PERCEPTION OF COGNITION – WE SEE CONNECTIONS

With schemata, we can internalize experiences,
we can make analogies, we can indulge
in the intricacies and pleasures
of higher-level thinking.

Schemata are the webs—the interconnectedness of our four underlying themes:

As all of these things become a part of us,
part of the way we think and act,
then we develop into whole, integrated human beings.

When we have a fully organized set of concepts,
which we are able to compare with the external world
during the PERCEPTION of COGNITION,
then we have moved beyond concepts to schemata.

SCHEMATA ARE DEVELOPMENTAL – WE EXPERIENCE WHOLE LEARNING

My schemata develop and change as I go through
the many stages of my growth.

As a young child, I am very concerned with self,
and early attitudinal and conceptual growth
will serve mainly to allow me to understand
my immediate world.

In the late primary years,
as I can more effectively relate worldly events to external causes,
my schemata take on a broader scope
and encompass more natural phenomena.

In my senior elementary years,
both my knowledge base and my conceptual understandings
are further enhanced by my developing maturity.
I am able to examine more far-reaching events
and recognize examples of inter-relatedness.

SCHEMATA TAKE US BEYOND CONCEPTS – WE ARE WHOLE PEOPLE

If my teacher approaches the instruction of Science
from the point of view of assisting in my development of schemata,
then integration is a simple and logical
extension of the program.

There are many ways in which the rest of the curriculum
is integrated into the basic Science program.

We brings Literature into the Science lesson,
and Science into Spelling.

We do theme units that combine
Research with Written Creativity, Art, Drama, and Music.

As I develop Scientific Schemata,
- or an integrated way of looking at the whole environment -
I can approach all my learnings
with the same Skills and Attitudes.

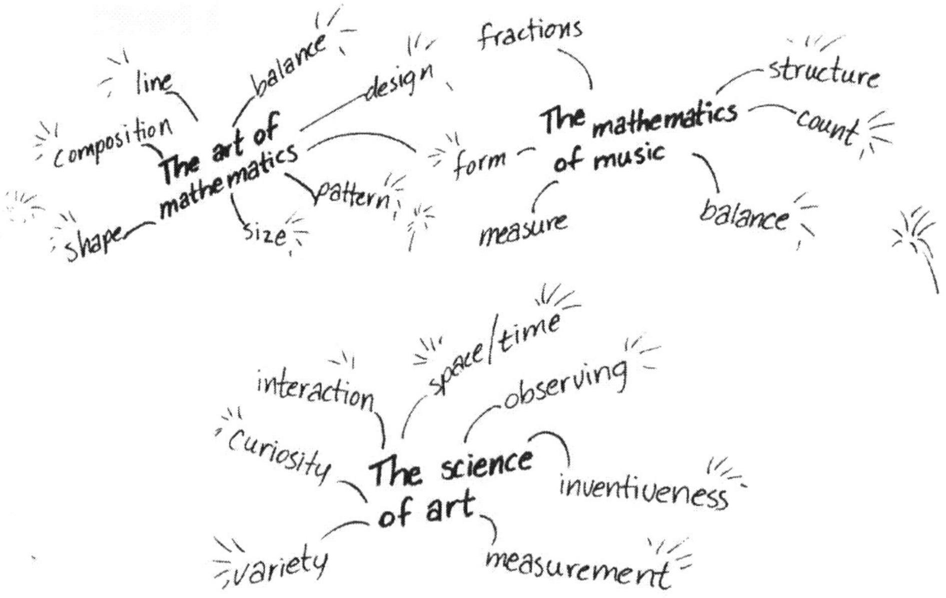

INTEGRATION

Using Themes

This is the most common approach to integration.

When using this method,
my teacher chooses a general Science topic,
and we brainstorm all the potential subtopics.

We then examine all aspects of the topic together.

This is often a content-based approach,
but it can involve much more,
as we expand our understanding of the principles.

(We devised this chart – and the list of Futuristics-Communications in 1987!)

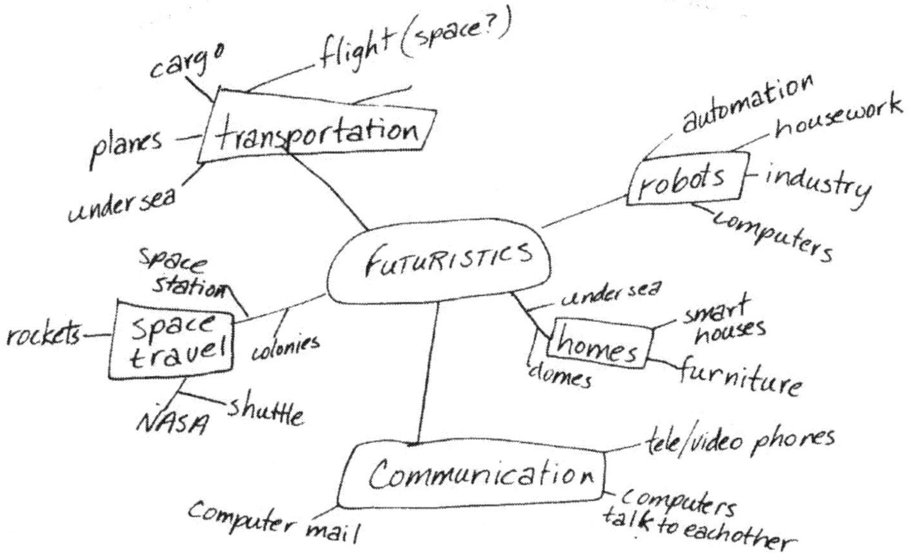

Using Projects

When using this method of integration,
my teacher sets up a variety of activities
that are related to the prime topic.

By using this technique,
we are forced to examine a single topic from many different angles,
consequently enhancing the growth of many thinking skills.

Some areas that the activities encompass are:

Written Creativity Research Drama
Artistic Creativity Model Building
Decision Making Planning Evaluation

In order to develop activities in many areas,
one has to "climb inside" the topic, and let one's mind wander.

My teacher uses the thinking skills
that we have been taught
and is able to find ways for us to look
at a specific topic from many different angles.

Where should spaceports be built? What effect would they have on the environment?		Design a travel guide for a trip through the Solar System.
	SPACE	
Draw a picture of life on a space station.		Design a spaceship that could go through Jupiter's atmosphere.

Using Specifics

A third method of integration can involve
the teaching of one specific concept, skill, or attitude
using more than one different subject area.

For example, if my teacher wants to teach me PERSISTENCE,
I will need more than just previously set-up scientific experiments
to perform, while my dedication to the task is being observed.

There might also be a sequence
of progressively more complicated math puzzles
that will encourage me to "stick with it".

Then we might examine examples of persistence
in the lives of explorers,
and in characters in our stories.

We could also create a play
that shows the effects of persistence
and draw pictures of situations
that demonstrate both the presence
and the lack of persistence.

And throughout the activities,
we constantly discuss the idea of persistence.

REMEMBER THE VALUE OF "DEBRIEFING".
WE ALL NEED TO SUMMARIZE OUR THOUGHTS AND FINDINGS
WHEN WE ARE LEARNING SOMETHING NEW.

SECTION 4 - DEVELOPING BASIC THINKING SKILLS

CREATIVITY

DIVERGENT THINKING

BRAINSTORMING

CONVERGENT THINKING

DECISION MAKING

FORECASTING

COMMUNICATION

PERCEIVING OPPORTUNITIES

TAKING ACTION

THINKING SKILLS – TO TRAIN OUR MINDS

THINKING SKILLS DEVELOP THOUGHTFUL SELF DIRECTION.

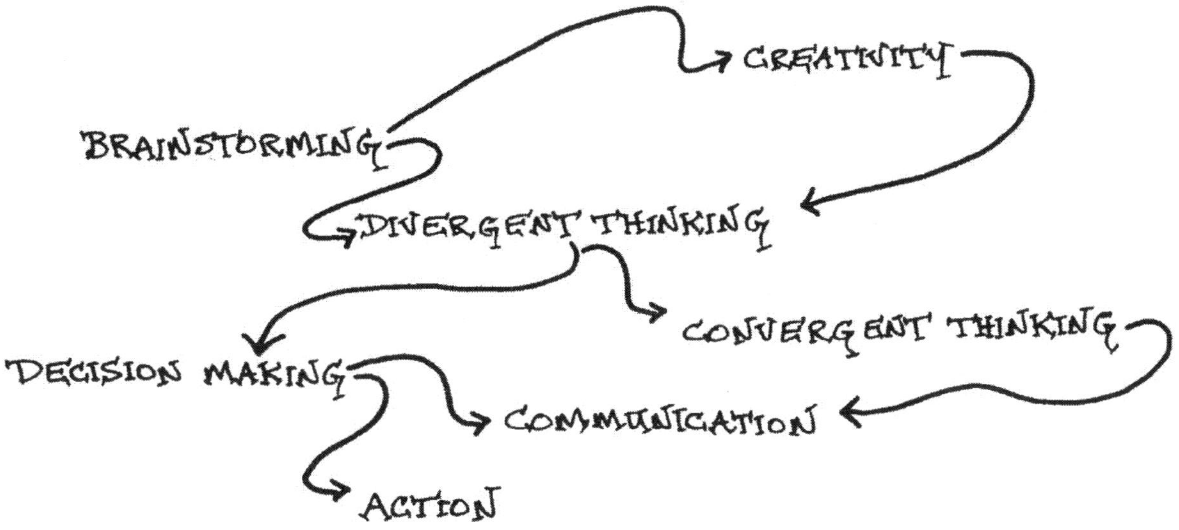

In order to get the most from my studies in science,
I need to develop a bank of thinking skills
to help me to develop thoughtful self-direction.

Creativity – We Think of Alternatives

CREATIVE TALENT

is the ability to go beyond the ordinary.

It involves putting together seemingly unrelated information to come up with new solutions, new ways of expressing ideas.

CREATIVITY INVOLVES ORIGINALITY.

- We consider the problem from unusual points of view.
- We relate previously unconnected facts and ideas to solve problems in new ways.
- We make unusual and unexpected responses.
- We give answers that no-one else has thought of.
- After sharing, we think of answers not yet considered.

CREATIVITY INVOLVES FLEXIBILITY.

- We consider the problem from many points of view.
- We find responses in many categories.
- We can list many ideas for a specific use or category.

CREATIVITY INVOLVES FLUENCY.

- We generate a superfluity of ideas.

DIVERGENT THINKING – WE EXPAND OUR THOUGHTS

DIVERGENT THINKING

is the generation of information
with the main emphasis on variety,
originality, and quantity of output
for each individual.

DIVERGENT THINKING

doesn't look for the best answer;
rather, it is concerned
with exploring ideas and their relationships.

DIVERGENT THINKING

is a higher level thinking process.

DIVERGENT THINKING INVOLVES ORIGINALITY.

➢ We offer original and unusual responses.
➢ We can be unique.
➢ We think of something no one else thought of.

DIVERGENT THINKING INVOLVES FLEXIBILITY.

➢ We look for variety.
➢ Our responses are free from categorical restrictions.

DIVERGENT THINKING INVOLVES FLUENCY.

➢ We generate a multitude of responses.
➢ All ideas are accepted and recorded.
➢ We ask for as many ideas as possible and encourage more than one response from all.
➢ We ask for many more ideas in a given category.
➢ We generate ideas.

BRAINSTORMING – WE GENERATE IDEAS

We like to start off each experimental session with a fit of brainstorming. That way we get our minds loosened up and start to consider all the possibilities.

The Rules...

NO JUDGEMENT

Any answer given during the session is accepted and recorded.

No member of the group is allowed to comment on someone else's suggestion.

Suggestions do not have to be defended.

THE WILDER THE BETTER

Allow enough time for responses to move beyond the ordinary.
Encourage wide-ranging thinking by accepting all responses unconditionally.

Keep asking, "What else?"

QUANTITY WANTED

The more answers given, the greater the creativity.

The more answers given,
the more likely that "the best" answer will be found,
if there is one.

The more answers given and accepted,
the less the children will fear offering new ideas.

PIGGYBACK O.K.

Answers can be similar to a previous answer, if there is a shade of difference.

Uses for a Straw

(brainstormed by a Grade 5 class)

to sip with	sword	bow	table legs
blow bubbles	toothpick	arrow	baseball bat
barber pole	swizzle stick	blades for a helicopter	hockey stick
pea shooter	guitar string	stir stick	money holder
earring	book marker	clothes line	needle
tunnel	clock hand	flag pole	bee swatter
hideout for ants	quiver	pencil holder	drum sticks
shoestring holder	poster holder	vacuum cleaner for plates	ruler

Convergent Thinking – We Contract Our Thoughts

CONVERGENT THINKING

is a process of selection.
The best alternative is chosen from many.

CONVERGENT THINKING

is definite, goal oriented thinking.
It involves making a single decision,
judgement or solution.

There does not need to be the same answer
from all participants,
but each answer must be a considered one.

CONVERGENT THINKING INVOLVES STATING THE PROBLEM.

- We are aware that a decision is necessary.
- We are able to define the exact area of concern.

CONVERGENT THINKING INVOLVES SELECTION.

- We make a decision or a choice.
- We state a preference.
- We explain how our decision meets the specifications.

CONVERGENT THINKING INVOLVES CONSIDERATION.

- We are concerned with considerations or restraints related to the problem.
- We are aware of related information.
- We are aware of goals and desired outcomes.

DECISION MAKING – WE REACH A CONCLUSION

DECISION MAKING

is the process of stating a preference or preferred action.

Yes, but REAL DECISION MAKING is not just choosing between two predetermined objects or circumstances.

It is the careful evaluation of data prior to making a preferred judgement.

DECISION MAKING INVOLVES EXPERIMENTAL EVALUATION.

- We see all the possible alternatives, and understand them.
- We recognize the existing limitations and conditions.
- We understand how our needs and wants are related to our decisions.

DECISION MAKING INVOLVES JUDGEMENT.

- We weigh the possibilities, have alternatives in mind, and act according to the decision.

DECISION MAKING INVOLVES LOGICAL EVALUATION.

- We list all possible alternatives and choose the best.
- We evaluate each from the list of best alternatives.
- We know the main goals and keep them in mind.
- We can support the decisions we make.
- We become aware of what can be accomplished as a result of the decision.
- We consider the best sequence to follow.

FORECASTING – WE LOOK TO THE FUTURE

FORECASTING

is the prediction or evaluation of future events.

FORECASTING

requires an ability to evaluate cause and effect,
with consideration given to a plan of action,
the possible effects of that action,
and anticipated consequences.

FORECASTING

looks at both the immediate future and at the distant future.

FORECASTING INVOLVES CONCEPTUAL AWARENESS.

- We know our goals.
- We are able to predict (based on observation and experience) the results of our actions.
- We think and plan before we act.

FORECASTING INVOLVES DETAILED CONSIDERATION.

- We analyze the details of our planned action.
- We make sure that we have all the information that we need to act sensibly.
- We examine ways in which a change in a situation may affect our plans or require a change in our approach.

FORECASTING INVOLVES SOCIAL CONSCIENCE.

- We are aware of others and ways in which our actions will affect them.
- We recognize possible destructive or harmful effects of our actions.
- We are sensitive to the needs of others.
- We act in a responsible way.

COMMUNICATION - WE DISCUSS OUR IDEAS

COMMUNICATION

is the cornerstone of human interaction.

COMMUNICATION

requires consideration of cultures,
traditions, mores, and individual goals.

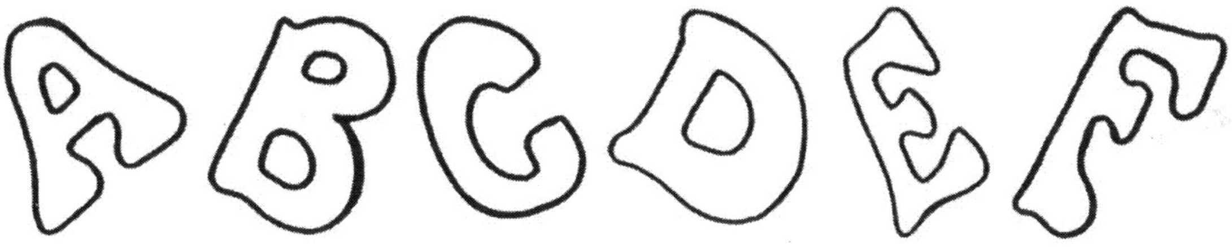

COMMUNICATION

requires consideration of
many variations in language usage,
in order to overcome barriers.

COMMUNICATION

helps us to create linkages with each other.
We are able to understand each other,
and therefore we can build relationships
based on trust and cooperation.

COMMUNICATION INVOLVES ACCURATE "SENDING" AND "RECEIVING."

- We are complete and specific in the information that we present.
- We understand non-verbal forms of communication, and use them carefully.
- We can express emotional reactions, and we can understand someone else's emotional reaction.

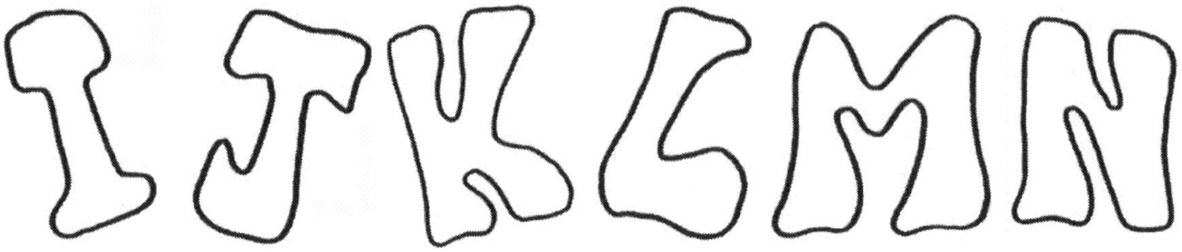

COMMUNICATION INVOLVES A KNOWLEDGE OF LANGUAGE.

- We use comparisons, analogies, similes, descriptive phrases, and other linguistic devices to present our ideas.
- We compare ideas and emotions to experience in order to establish common ground with others.
- We understand the interrelatedness of language and experience.
- We use words to add depth of meaning to our message.

COMMUNICATION INVOLVES SOCIAL CONSCIOUSNESS.

- We use appropriate language to add humor, show concern, and express values in our communication.
- We acknowledge the values of others and try to eliminate any usage that may cause stress.

Perceiving Opportunities - We Are Aware of Options

PERCEIVING OPPORTUNITIES

is the ability to discern ways
of exploiting any given set of conditions
in order to realize personal gain.

PERCEIVING OPPORTUNITIES

allows us to develop personally, intellectually, and socially.

PERCEIVING OPPORTUNITIES

is a life skill,
and involves an awareness of the past, the present, and the future.

PERCEIVING OPPORTUNITIES INVOLVES FLEXIBILITY.

- We set realistic goals.
- We are flexible in our choice of alternatives.
- We plan for a variety of potential outcomes.

PERCEIVING OPPORTUNITIES INVOLVES DETERMINATION.

- We constantly look for opportunities.
- We make sure that we have done all the preliminaries before we proceed.
- We are aware of our own potential.
- We do not let the possibility of disappointments or setbacks deter us.

PERCEIVING OPPORTUNITIES INVOLVES JUDGEMENT.

- We analyze situations to ensure that our choices are rational.
- We recognize possible roadblocks to success and develop ways of overcoming them.
- We do not look to take opportunities that may hurt others or the environment.

TAKING ACTION - WE ARE DECISIVE

TAKING ACTION

is the culmination
of the entire process of problem solving.

TAKING ACTION

requires a focus,
which comes from a logical analysis
of the situation.

TAKING ACTION

is a step beyond decision making,
for it IS action,

and not just a mental activity.

TAKING ACTION INVOLVES INTEGRATION.

➢ We develop strategies for using all our thinking skills.

➢ We can take a broad view.

TAKING ACTION INVOLVES SOCIAL CONSCIENCE.

➢ We do not infringe on the rights and goals of others.

➢ We recognize that if we limit the rights of others, then ultimately we limit our own rights.

TAKING ACTION INVOLVES LONG RANGE PLANNING.

➢ We look to the future and consider both the benefits and the potential pitfalls of our actions.

SECTION 5A - EXAMINING BASIC PROCESS SKILLS

OBSERVATION

INFERENCE

COMMUNICATIONS

MEASUREMENT

CLASSIFICATION

SPACE / TIME RELATIONSHIPS

BASIC SKILLS - TO PERFORM WITH

SKILLS HELP US LEARN HOW TO SOLVE PROBLEMS IN A LOGICAL WAY.

All of these skills need to be developed
and reinforced sequentially
throughout the science program.

My teacher doesn't plan lessons
just to teach us a particular skill.

We have activity sessions
in which we can use a number of different skills
as we look at the world around us
and try to solve different problems.

Sometimes we get a lesson that helps us learn just one skill.
The teacher always lets us know what
we're trying to learn as we do our activities.

OBSERVATION – WE SEE

OBSERVATION IS PERCEPTION – USING ANY OF THE SENSES.

Demonstrating the Skill

I TRY TO:

➥ use all my senses (if it is safe to do so) when making observations.

➥ make accurate measurements of size, number, mass (quantitative observations).

➥ extend the range of my senses using various instruments. (e.g., magnifiers, measuring instruments, etc.)

➥ look for change, and make careful observations of any changes that I notice.

➥ be aware of the differences between observations and inferences.

Developing the Skill

OBSERVATION IS THE BASIS OF ALL INQUIRY SKILLS.

Familiar materials
can be the start
of many valuable
OBSERVATION activities.

We like to try
such things as
lighting candles,
bouncing balls,
watching tadpoles grow.

Some properties
may stay the same,

some properties
may change,

and some properties
that we did not previously notice
may appear
when we examine things more closely.

OBSERVATION IS THE PROCESS BY WHICH I CAN OBTAIN DIRECT INFORMATION ABOUT MY SURROUNDINGS.

Evaluation of the Skill

IN GRADES 1-3, I SHOULD BE ABLE TO:

- utilize all of my senses as I gather information.
- describe objects I examine by referring to the information I have gained.
- recognize a variety of objects or organisms from a written or oral description.
- distinguish between several different objects on the basis of color, size, shape, temperature, texture, taste, smell, etc.
- observe many objects in great detail.
- notice changes in objects, events, and organisms.

BY GRADES 4-7, I CAN:

- recognize those observations that are relevant to a particular investigation and those that are not.
- recognize the differences between causes and effects.

Summative Evaluation

LEVEL 1: At a SUPERIOR level, I make more observations; I can determine the more subtle aspects of an observation; I can make finer distinctions; I analyze more intensely; I consider more possibilities before making my conclusions.

LEVEL 2: On the AVERAGE, I would fulfill the assignment as given; make several observations; recognize general aspects or changes; examine several possibilities.

LEVEL 3: My work is SATISFACTORY if I work at the task given, using all my senses when reminded; make some observations; recognize broad aspects of change; determine somewhat more than the fundamental attributes.

LEVEL 4: My work would be UNSATISFACTORY if my examination of an object or event is frivolous; if I make few observations; if I guess at identification; if I recognize only the most obvious of attributes.

SO, IF MY TEACHER IS GOING TO EVALUATE MY ABILITY TO MAKE OBSERVATIONS, THAT IS WHAT SHE LOOKS FOR!

MAKING INFERENCES – WE UNDERSTAND

INFERRING IS THE PROCESS BY WHICH WE EXPLAIN OR INTERPRET AN OBSERVATION.

More than one INFERENCE can be made
from a single observation, though.
And so, you see,
an INFERENCE may be logical
and yet be wrong!

Developing the Skill

THE MORE OBSERVATIONS THAT SUPPORT AN INFERENCE, THE MORE LIKELY IT IS TO BE CORRECT.

so...
my teacher provides
further observations to help
me and the other children
as we try to support
our inferences.

AN INFERENCE COMBINES PAST KNOWLEDGE WITH NEW INFORMATION.

Evaluation of the Skill

IN GRADES 1-3, I SHOULD BE ABLE TO:

- suggest a reasonable explanation based on the evidence of what I see.
- suggest reasonable explanations for different observations (e.g., why an animal behaves as it does).
- tell the difference between statements which are observations and those which are inferences.
  Observation: this liquid looks like water
  Inference: this liquid is water
- suggest a variety of inferences when I explain an observation or an event.

BY GRADES 4-7, I CAN:

- distinguish between those observations which can reasonably be used to support an inference and those which cannot.
- distinguish between an inference which fits all the observations and one which does not.

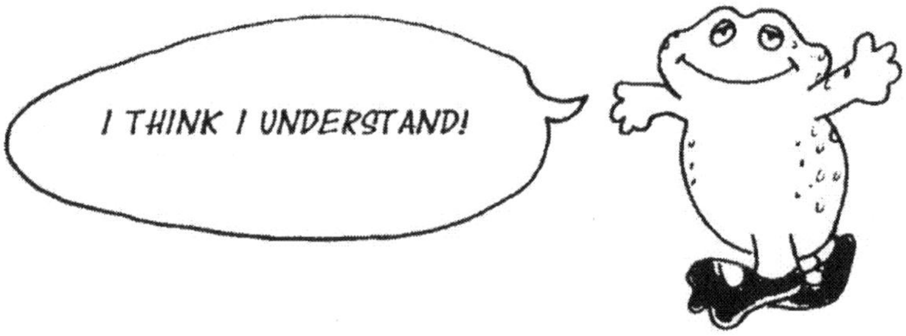

Summative Evaluation

LEVEL 1: At a SUPERIOR level, I can see and express a multiplicity of possibilities and then use my observations to determine the most reasonable alternative.

LEVEL 2: On the AVERAGE, I can recognize several possibilities and suggest explanations that are congruent with my reasoning.

LEVEL 3: My work is SATISFACTORY if I realize that there is more than one inference to fit an observation, and I can offer a satisfactory explanation of events.

LEVEL 4: My work would be UNSATISFACTORY if my inferences are guesses or they do not fit the observations, or if my explanations are limited.

SO, IF MY TEACHER IS GOING TO EVALUATE MY ABILITY TO MAKE INFERENCES, THAT IS WHAT HE LOOKS FOR!

COMMUNICATION - WE EXPLAIN

COMMUNICATION IS THE PROCESS BY WHICH WE EXPLAIN OR INTERPRET AN OBSERVATION.

Developing the Skill

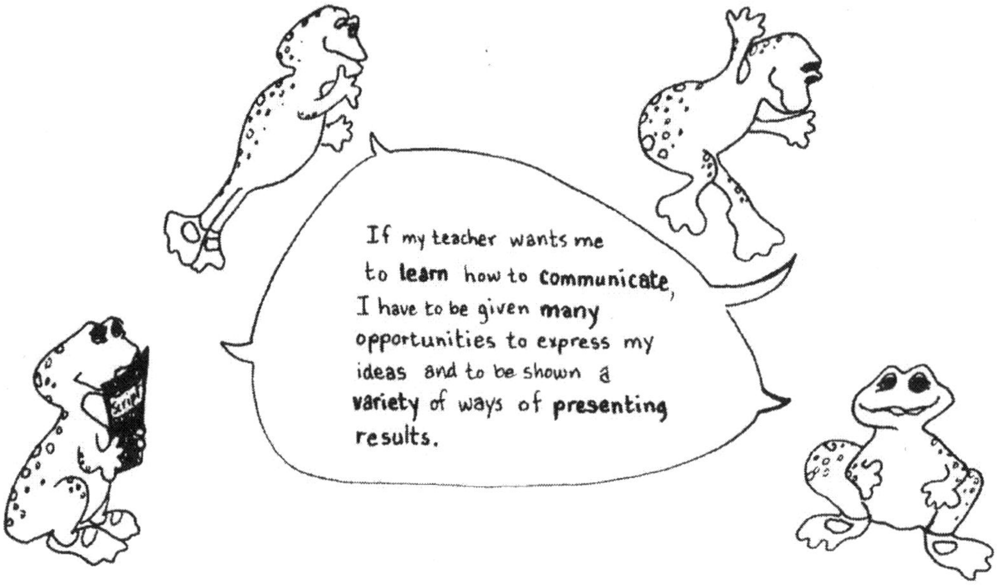

COMMUNICATION HELPS CHILDREN TO CLARIFY THEIR OWN IDEAS, AS WELL AS TO LEARN FROM OTHERS.

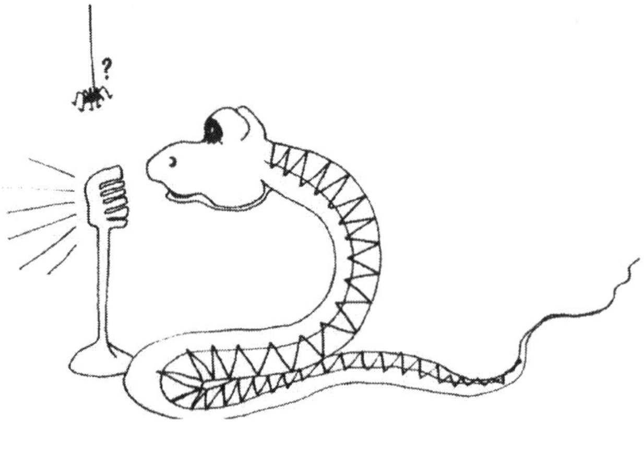

Types of Communication

Evaluation of the Skill

IN GRADES 1-3, I SHOULD BE ABLE TO...

- clearly describe an object, event, or organism in simple and general terms.
- describe some of the similarities and differences among objects and events.
- describe an object so that it can be identified by my classmates.
- determine if the information given is sufficient to enable a definite identification of an object.
- describe the changes that take place in an object or a living thing.

BY GRADES 4-7, I CAN...

- construct a variety of data tables and graphs.
- select relevant and important information for communication.
- select a method of presenting information which takes into account the content, my audience, and my purpose.
- sketch objects and organisms to scale.

Summative Evaluation

LEVEL 1: At a SUPERIOR level, I describe observations clearly, noting subtleties in similarities and differences; recognize ambiguities; choose the most appropriate method of recording.

LEVEL 2: On the AVERAGE, I can describe observations; notice general similarities and differences; sort out major ambiguities; know and use several different methods of recording information.

LEVEL 3: My work is SATISFACTORY if I can describe an event or object in basic terms; avoid gross ambiguities; correctly use recording methods that are suggested to me by my teacher.

LEVEL 4: My work would be UNSATISFACTORY if my descriptions are vague and incomplete; if I ignore differences between objects; if I am unable to present my information in a rudimentary form.

SO, IF MY TEACHER IS GOING TO EVALUATE MY ABILITY TO COMMUNICATE, THAT IS WHAT SHE LOOKS FOR!

MEASUREMENT – WE COMPARE TO A STANDARD

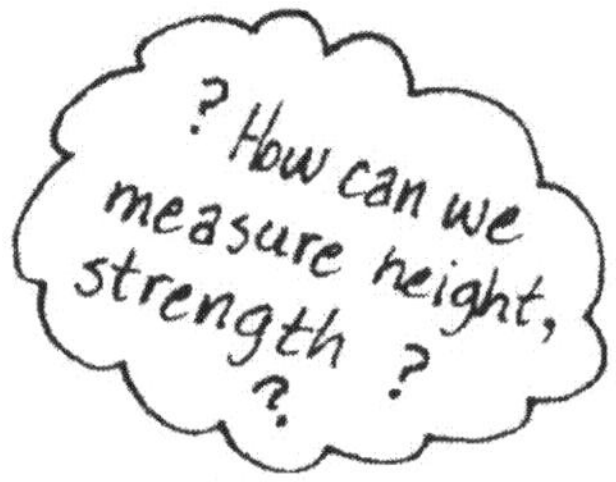

MEASUREMENT
IS THE PROCESS
OF COMPARING A PROPERTY...

TO A STANDARD...

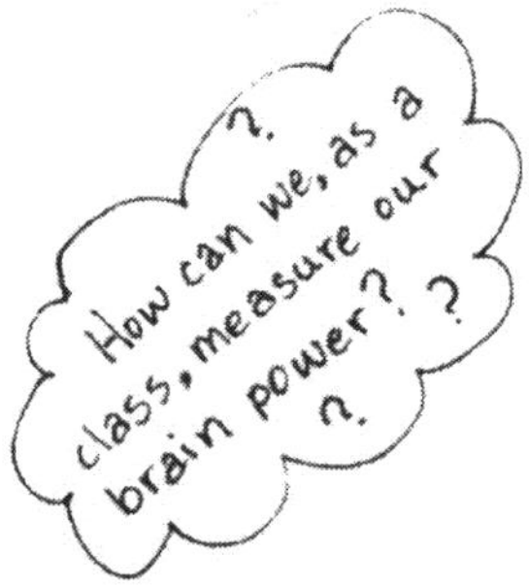

SO THAT
OBSERVATIONS...

CAN BE DESCRIBED...

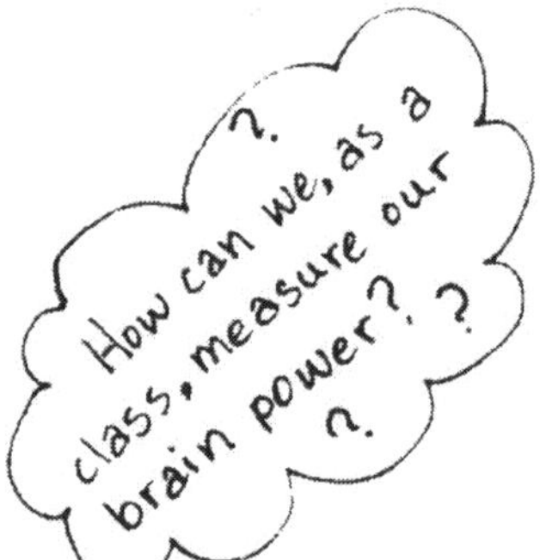

MORE ACCURATELY.

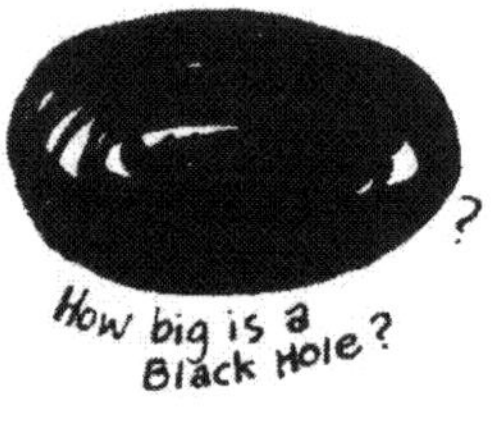

Developing the Skill

WE LEARN HOW TO MEASURE BY...

➥ making simple comparisons using non-standardized units. (E.g., cut lengths of string or paper to show the circumference of apples, and use these measurements to describe one apple as "fatter" or "thinner" than the others.)

➥ comparing an object to a standard item (e.g., a pen has the weight of 15 thumb tacks, a pencil only 10).

➥ learning about standardized units of measurement.

WE CAN MEASURE LOTS OF THINGS USING LOTS OF TOOLS...

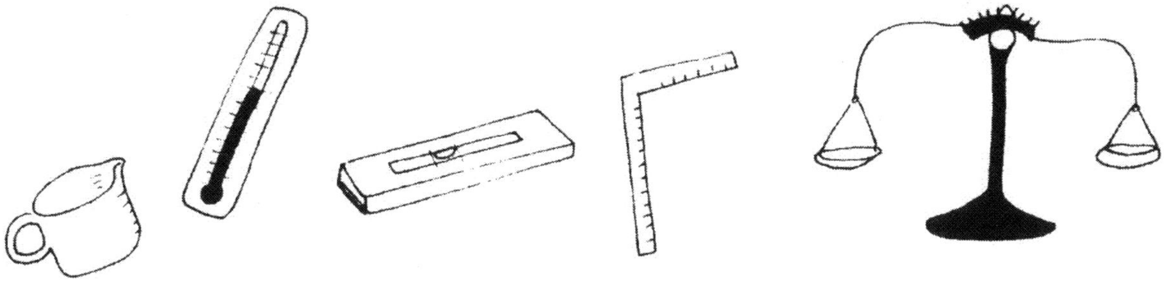

Evaluation of the Skill

IN GRADES 1-3, I SHOULD BE ABLE TO...

- measure and compare height, weight, and length.
- compare the sizes of various parts of the body.
- do patterning and sequencing activities.
- understand how to graduate size (e.g., beakers, rulers).
- compare objects by size (e.g., taller than).
- sequence events, using words such as before, during, after, first, second, next, etc.

BY GRADES 4-7, I CAN...

- compare quantities or sizes with reasonable efficiency.
- measure accurately enough to make comparisons between objects and events.
- select the proper units and instruments for measuring different events, objects, and organisms.

Summative Evaluation

LEVEL 1: At a SUPERIOR level, I understand intricate patterns and sequences; I correctly choose units and instruments for measurements of any kind; I make comparisons of quantity or size as a regular part of my classroom science work.

LEVEL 2: On the AVERAGE, I am able to identify many different patterns or sequences; I measure with care using my instruments carefully; I usually select the correct units and equipment.

LEVEL 3: My work is SATISFACTORY if I can measure items accurately when given the units and instruments; I choose the correct units from a limited choice; I recognize general patterns and broad sequences.

LEVEL 4: My work would be UNSATISFACTORY if I am unable to make valid comparisons; I cannot follow more than the most basic of sequences; I make no effort to make measurements during experiments; I cannot make accurate measurements; I cannot choose appropriate units for measurement.

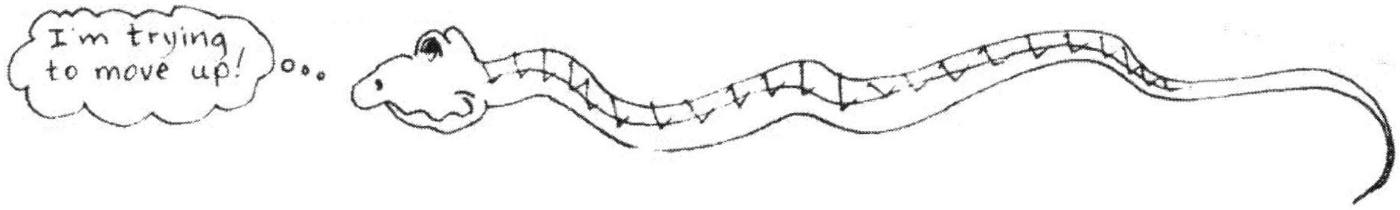

SO, IF MY TEACHER IS GOING TO EVALUATE MY ABILITY TO MAKE MEASUREMENTS, THAT IS WHAT HE LOOKS FOR!

CLASSIFICATION ~ WE SEE PATTERNS

CLASSIFICATION
IS THE PROCESS BY WHICH OBJECTS AND EVENTS ARE GROUPED ACCORDING TO COMMON PROPERTIES.

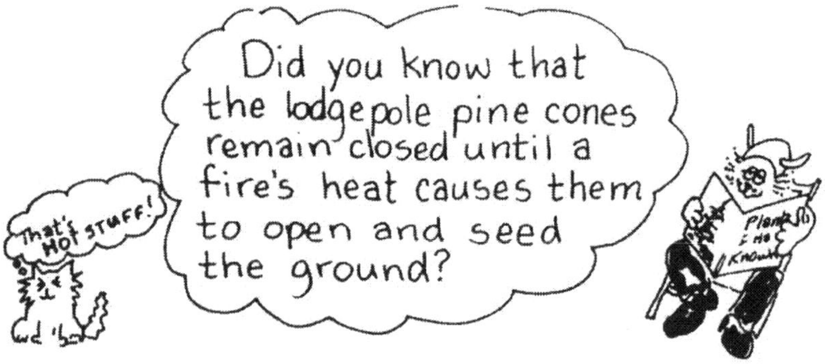

Types of Classification

CLASSIFICATION ENABLES US TO ORGANIZE THE WIDE VARIETY OF THINGS AROUND US INTO CONVENIENT GROUPS.

The CLASSIFICATION that I use is determined by my purpose for doing the grouping and by my personal preferences.

e.g., grouping leaves:
- according to size or shape or color
- their position on a twig
- the pattern of their veins

Developing the Skill

I CAN DEVELOP MY ABILITY TO CLASSIFY BY...

- ➥ careful observation of objects and events.
- ➥ identification of the similarities and differences in objects and events.
- ➥ trying to develop my own classification schemes.
- ➥ including an explanation of the methods I've used.
- ➥ finding alternative ways of classifying things.

Evaluation of the Skill

IN GRADES 1-3, I SHOULD BE ABLE TO...

- use a single property or attribute to arrange a set of objects into two groups.
- sort, match, group, and name various colors and objects.
- recognize a single, common property in a set of objects.
- use one common characteristic to group a set of objects into two or more groups.

BY GRADES 4-7, I CAN...

- identify similarities and differences among objects, events, and organisms.
- identify common structural or behavioral patterns in objects, events, and organisms.
- read simple classification schemes and identification keys, and construct such keys.

Summative Evaluation

LEVEL 1: At a SUPERIOR level, I can make subtle distinctions as to the properties of objects; I find many similarities and differences when examining objects and situations; I prepare well-defined classification keys.

LEVEL 2: On the AVERAGE, I can group objects based on two or more common properties; I can recognize several similarities and differences in objects or situations; I can prepare and read classification keys.

LEVEL 3: My work is SATISFACTORY if I can group objects based on a single property; I recognize the major similarities and differences in objects or situations; I read and prepare simple classification keys.

LEVEL 4: My work would be UNSATISFACTORY if I cannot group objects according to even simple properties; I recognize only the most obvious of differences or similarities; I cannot read or construct even a simple classification key.

SO, IF MY TEACHER IS GOING TO EVALUATE MY ABILITY TO CLASSIFY, THAT IS WHAT SHE LOOKS FOR!

SPACE / TIME RELATIONSHIPS – WE STUDY ORDER AND CHANGE

WHEN USING *SPACE AND TIME RELATIONSHIPS* WE ARE CONCERNED WITH DEVELOPING AN UNDERSTANDING OF OBJECTS IN TERMS OF THEIR SIZE, SHAPE AND POSITION, AND THE CHANGES IN THOSE OVER A CERTAIN PERIOD OF TIME.

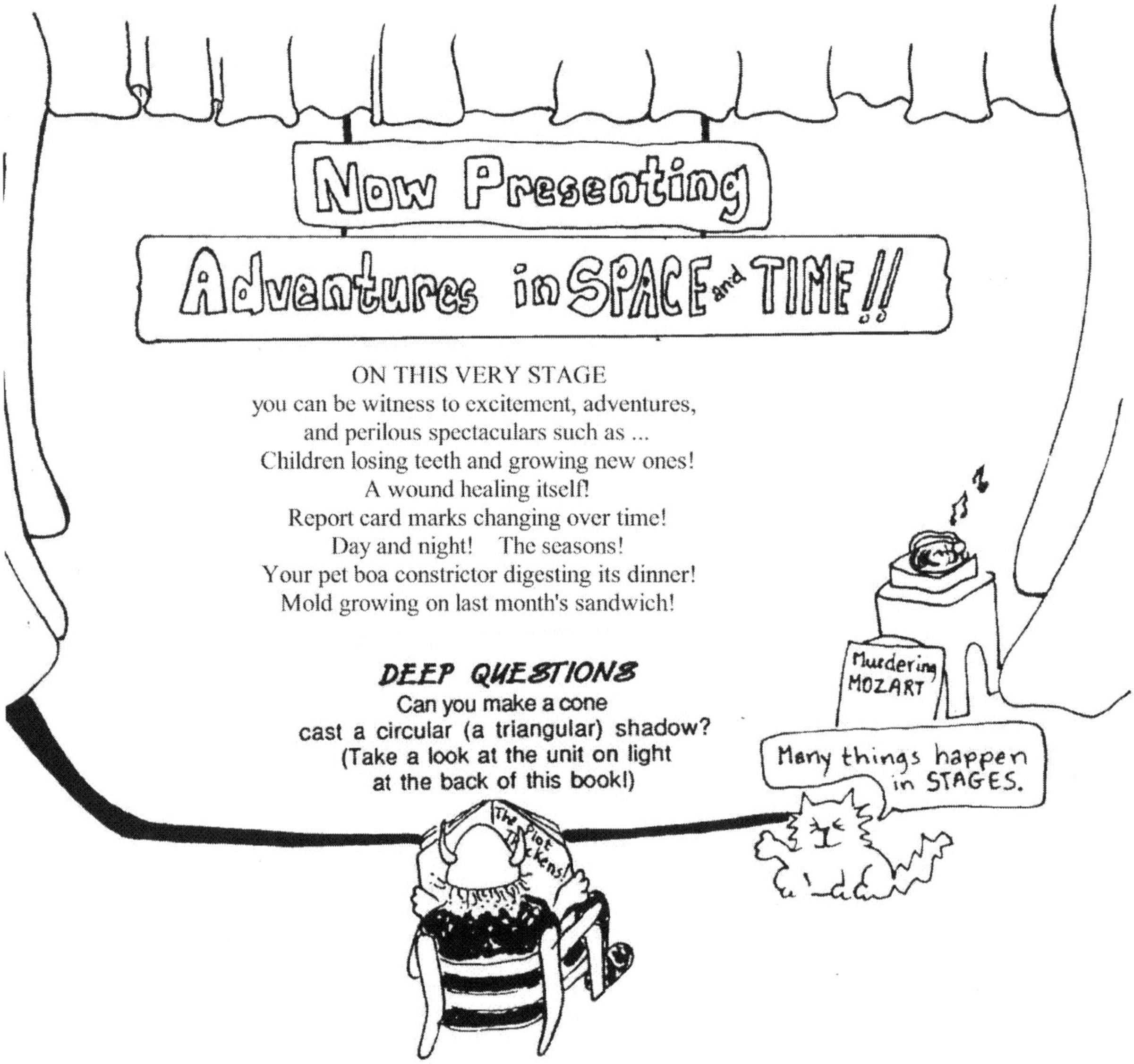

Developing the Skill

WE CAN DEVELOP AN UNDERSTANDING OF THESE RELATIONSHIPS BY...

- observing a variety of different shapes.
- observing how an object's shape may appear to change when it is viewed from different positions.
- observing two-dimensional representations of objects (in silhouettes or in shadows).
- sequencing objects according to position, size, and shape.
- sequencing events in time.

Evaluation of the Skill

IN GRADES 1-3, I SHOULD BE ABLE TO...

- sequence and classify objects by size and location.
- sequence events in time.
- accurately use such words that describe space and time. (e.g., up/down, right/left, forward/backward, toward/away from, above/below, earlier/later, soon/eventually, etc.)
- recognize two-dimensional shapes that are components of three-dimensional shapes. (e.g., the circular bottom of a cone, its triangular side, etc.)

BY GRADES 4-7, I CAN...

- recognize symmetry in objects, events, and organisms.
- compare events in terms of rate of growth, movement, life cycles, etc.
- ➤ describe the positions and appearances of various objects relative to myself and to other observers.

Summative Evaluation

LEVEL 1: At a SUPERIOR level, I would understand subtleties in size, shape, and location; I recognize symmetry in a wide variety of objects; I sequence events within narrow boundaries; I fully understand and correctly use words that describe location and relative time.

LEVEL 2: On the AVERAGE, I would understand several differences in size, shape, and location; I recognize symmetry in many patterns and designs; I am able to generally sequence events; I generally use and understand words that describe location and relative time.

LEVEL 3: My work is SATISFACTORY if I can recognize broad differences in size, shape and location; I recognize symmetry in broad patterns; I sequence events with the help of cue words; I understand time-sequencing words on a broad scale.

LEVEL 4: My work would be UNSATISFACTORY if I cannot recognize symmetry in simple patterns; I cannot sequence events even with several cues; I cannot explain broad differences in size, shape, and location.

SO, IF MY TEACHER IS GOING TO EVALUATE MY UNDERSTANDING OFSPACE-TIME RELATIONSHIPS, THAT IS WHAT HE LOOKS FOR!

SECTION 5B - EXAMINING ADVANCED PROCESS SKILLS

PREDICTING

CONTROLLING VARIABLES

INTERPRETING DATA

MAKING OPERATIONAL DEFINITIONS

FORMULATING HYPOTHESES

DESIGNING EXPERIMENTS

NOTE FOR THE TEACHER

Even though this section presents higher-level skills
than those examined in Section 5A,
they CAN and SHOULD be introduced
at the primary level—yes, even Grade 1!

They are still extremely important in the growth
of a basic understanding of the processes of Elementary Science.
Full understanding will come as the whole range
of skills is developed over the years.
Like the Basic Skills, they need to be
developed and reinforced sequentially
throughout the Science Program.

We just keep plugging away!

ADVANCED SKILLS - TO REFINE OUR ABILITIES

PREDICTING - WE LOOK TO THE FUTURE

The Value of Prediction

Through the process of prediction,
we force ourselves to examine outcomes
and accept that change might be necessary.

We learn how to examine these outcomes
with a deepened sense of personal responsibility.

Personal growth is thus greatly enhanced
on the road to becoming
an earth-conscious citizen.

When we put all of this together,
we grow—emotionally and intellectually.

The interrelation of skills, attitudes, and concepts
becomes obvious and essential!

The Basis of Prediction

PREDICTIONS MAY BE BASED ON...

Developing the Skill

WE CAN DEVELOP THE SKILL OF PREDICTING BY...

➥ observing regular events.

➥ analyzing individual components of an event.

➥ comparing current events with past experiences.

➥ noting many differing variables.

➥ examining all the possibilities.

➥ testing several predictions.

➥ changing some of the variables.

➥ graphing quantitative information, then making predictions from our graphs.

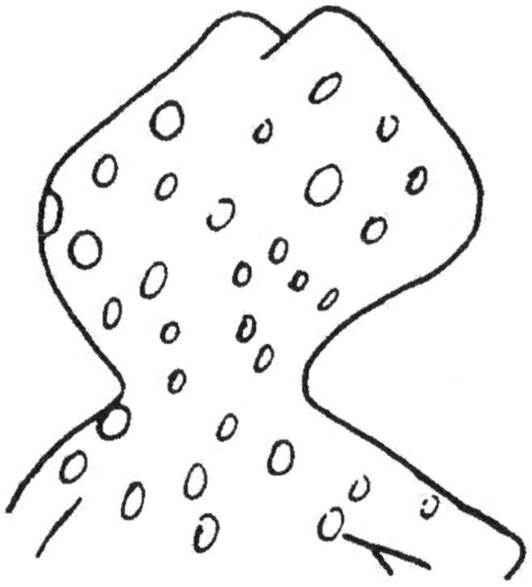

Evaluation of the Skill

IN GRADES 1-3, I SHOULD BE ABLE TO...

- begin to predict the possible outcome of events and actions.
- suggest what may happen as a result of simple actions.
- suggest what comes next in a series, in a story, in a game.
- discuss natural cycles and patterns.

BY GRADES 4-7, I CAN...

- anticipate problems, safety factors, etc. when conducting an investigation.
- listing several variables to be considered.
- recognize developing patterns.
- examine sensory input.
- use graphs and other patterns of data to make predictions.

Summative Evaluation

LEVEL 1: At a SUPERIOR level, I am able to examine a fairly extensive series of events and make a logical and generally accurate prediction as to the subsequent events; I am able to examine many different life cycles and predict changes in a particular case; I always recognize safety considerations; I am able to use a wide variety of graphs and tables to help in my predictions.

LEVEL 2: On the AVERAGE, I can examine several events in a series and predict the outcome; I can recognize the patterns in several life cycles; I respect and recognize safety considerations in experimentation; I am able to use several graphs and tables satisfactorily.

LEVEL 3: My work is SATISFACTORY if I can follow a regular classroom experiment and predict the next event or outcome; I can discuss the life cycle of an animal under observation; I follow classroom safety regulations; I can duplicate a given graph or table.

LEVEL 4: My work would be UNSATISFACTORY if I fail to make any logical predictions following a series of observations; I cannot describe life cycles that have been studied; I make tables and graphs that convey little information.

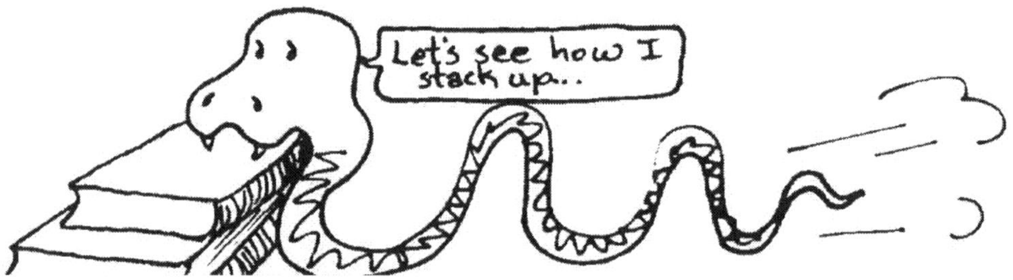

SO, IF MY TEACHER IS GOING TO EVALUATE MY ABILITY TO PREDICT, THAT IS WHAT SHE LOOKS FOR!

Controlling Variables - We Are in Charge

CONTROLLING VARIABLES IS THE PROCESS BY WHICH CONDITIONS AFFECTING AN EVENT ARE DELIBERATELY CONTROLLED AND MANIPULATED.

In a test,
all factors other than the one being tested must be held constant,
in order for the test to be considered fair.

For example,
if the children are testing which of two objects will roll further,
they need to be aware that many things will affect
the distance rolled:
the angle of the ramp;
the position on the ramp;
the surface onto which the objects roll, and so on.

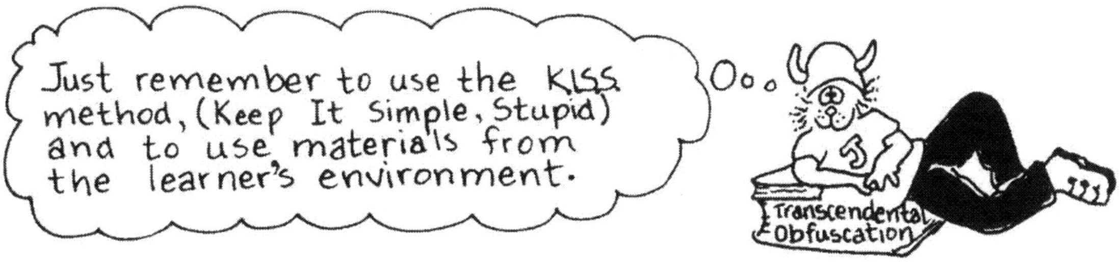

Developing the Skill

I CAN DEVELOP MY ABILITY TO CONTROL VARIABLES BY...

➥ differentiating between attributes and variables.

➥ learning to identify all the variables involved.

➥ changing variables in a logical sequence.

➥ observing the results of changing the selected variables.

➥ setting up single and multiple variable activities.

➥ devising tests in which one variable is changed while others stay the same.

➥ recording any changes and the effect of the change.

Examples of Variables

Evaluation of the Skill

IN GRADES 1-3, I SHOULD BE ABLE TO...

- examine all the steps of an experiment.
- identify the major steps of a comparison.
- tell the difference between simple attributes and variables.
- identify components of an investigation that make a comparison unfair (e.g., comparing the height of one object with the weight of another).
-

BY GRADES 4-7, I CAN...

- identify a group of factors which may have caused a change in an object or organism.
- investigate the effect of one variable at a time, by isolating single variables.
- specify which variables in an investigation are to be controlled, and which one is to be investigated.
- develop a variety of ways of controlling variables in an experimental situation.

Summative Evaluation

LEVEL 1: At a SUPERIOR level, I can identify the variables in complex experiments; I never change more than one variable at a time; I can devise ways to record effectively the results of all stages of my experiments.

LEVEL 2: On the AVERAGE, I can identify several variables; I generally remember to change only one variable at a time; I can use standard techniques to record experimental results.

LEVEL 3: My work is SATISFACTORY if I am able to distinguish between attributes and variables; if I examine one variable at a time when shown how to; if I can fill in record charts that are provided with the activity.

LEVEL 4: My work would be UNSATISFACTORY if I am unaware of what a variable is; if I constantly change variables while doing an experiment; if I am unable to keep simple records.

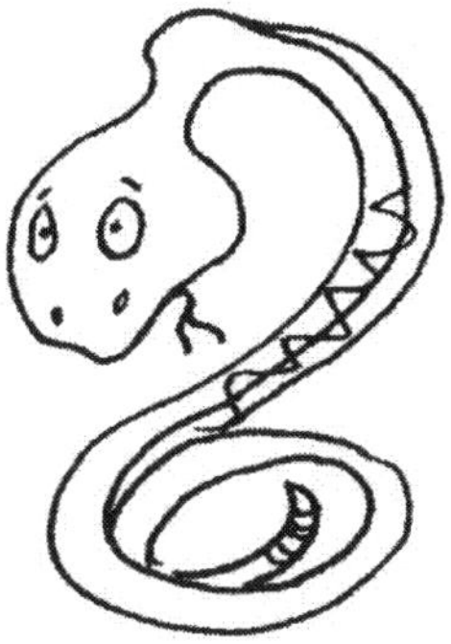

SO, IF MY TEACHER IS GOING TO EVALUATE MY ABILITY TO CONTROL VARIABLES, THAT IS WHAT HE LOOKS FOR!

Interpreting Data - We Study Our Findings

INTERPRETING DATA IS THE PROCESS BY WHICH WE ORGANIZE AND PRESENT DATA.

It is important that as much as possible is learned from the results of experimental observations and measurements.

Charts, graphs, tables, and diagrams may be used to record and to explain the results of experiments and investigations.

These can then be used by others as they try to understand what the experiment has shown.

Developing the Skill

I CAN DEVELOP MY ABILITY TO INTERPRET DATA BY...

- identifying the items being investigated.
- identifying the variables in the investigation.
- identifying the relationships between two or more variables.
- identifying any patterns and trends in data that describe the results of an experiment.
- identifying the direction and extent of any changes that can be observed in objects and organisms.
- identifying the underlying purpose of the investigation.
- drawing logical conclusions from the data.

Meanwhile, back at the lab, Professor Data was busy drawing a **logical conclusion.**

Examples of Data

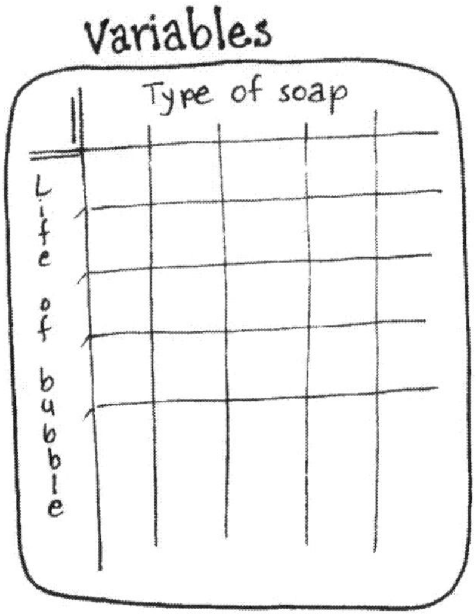

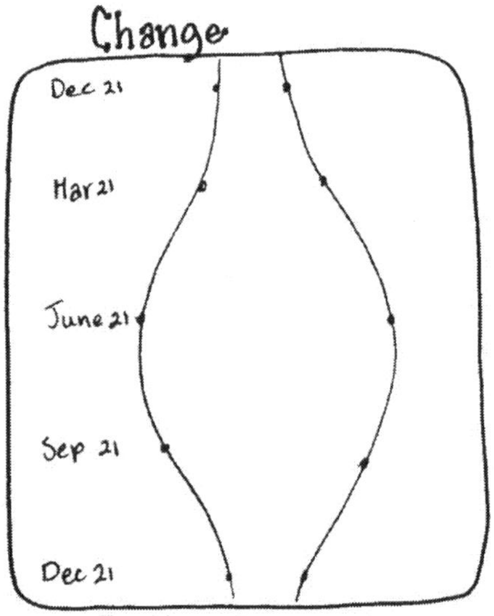

Sunrise - Sunset Times

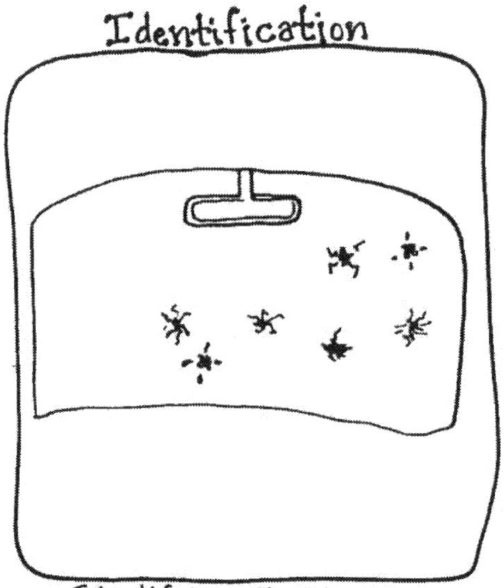

Identifying Windshield Organisms

Evaluation of the Skill

IN GRADES 1-3, I SHOULD BE ABLE TO...

- give explanations that are consistent with my observations.
- identify a variety of objects or organisms from a description of the characteristics.
- classify various objects or events according to their common characteristics.

BY GRADES 4-7, I CAN...

- read a variety of charts, tables, graphs, keys, etc.
- identify objects, substances, organisms, events, and so on from previously recorded data.
- recognize and explain patterns, relationships, and inconsistencies in data.
- use data collected to help me as I make predictions and as I challenge or construct hypotheses.

Summative Evaluation

LEVEL 1: At a SUPERIOR level, my explanations are always consistent with my observations; I am able to read complex graphs and tables; I spot inconsistencies readily; I can use data to make reasoned predictions.

LEVEL 2: On the AVERAGE, I make reasonable explanations and predictions based on my observations; I can interpret several kinds of graphs, tables, keys, etc.; I can recognize inconsistencies that are demonstrated to me.

LEVEL 3: My work is SATISFACTORY if I am willing to make explanations and predictions from my data; if I can read basic graphs and keys; if I can recognize blatant inconsistencies.

LEVEL 4: My work would be UNSATISFACTORY if I am unable to make even simple explanations or predictions; if I cannot spot even the most obvious of inconsistencies; if I am unable to extract information from simple graphs and tables.

SO, IF MY TEACHER IS GOING TO EVALUATE MY ABILITY TO INTERPRET DATA, THAT IS WHAT SHE LOOKS FOR!

Making Operational Definitions – We Know the Rules

MAKING OPERATIONAL DEFINITIONS
IS THE PROCESS BY WHICH
A SCIENTIFIC TERM IS DEFINED
ACCORDING TO WHAT MUST BE DONE
AND WHAT SHOULD BE OBSERVED
IN ORDER TO IDENTIFY THE CONCEPT.

For Example...

the strength of several cylinders,
each made of different materials,
could be tested in a variety of ways.

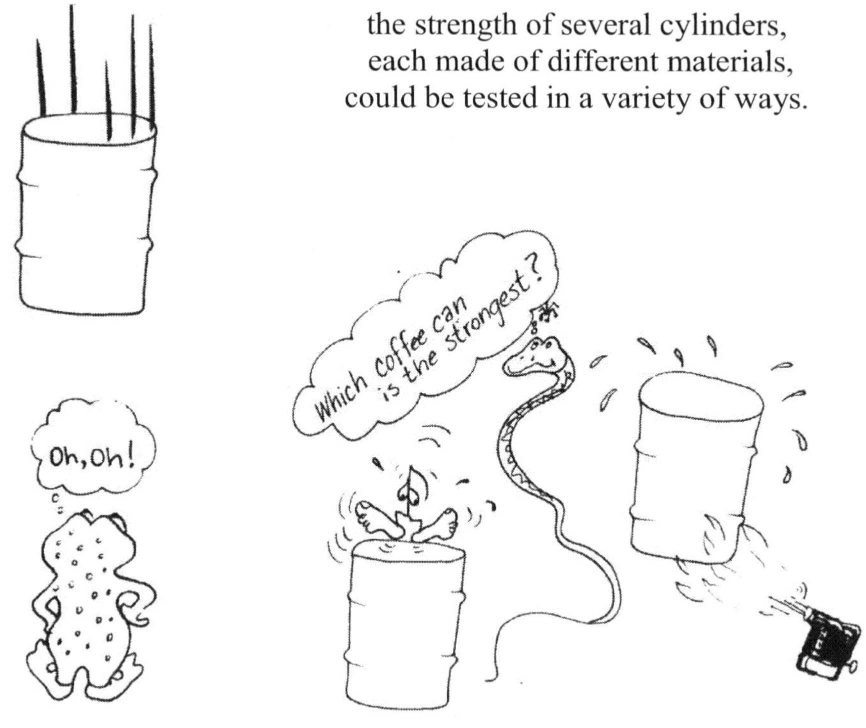

They could be struck by a heavy weight, dropped onto a hard surface, or exposed to great heat.

However, if we decide to define strength as:

"the weight the cylinder can withstand
without crumpling or cracking
when standing on its base,"

then we can make meaningful comparisons
as we test each of the cylinders.

Developing the Skill

I CAN DEVELOP MY ABILITY TO MAKE OPERATIONAL DEFINITIONS BY...

➥ realizing that a property that appears to be simple (e.g., strength, elasticity, hardness) can be interpreted in many different ways.

➥ showing that there are various ways of testing ideas and a variety of instruments available for making observations.

➥ recognizing the need to standardize measurements, and learning how to be accurate in my own measurements.

➥ realizing and demonstrating that the definition used will affect the results obtained.

Evaluation of the Skill

IN GRADES 1-3, I SHOULD BE ABLE TO...

- say what I mean when I describe an object or event.
- recognize that there is ambiguity in the use of the word "best" and understand that I need some rules that I can use when I am selecting the "best" paper plane, the "best" bridge, etc.
- recognize the different attributes of items that I use when I am doing my experiments.

BY GRADES 4-7, I CAN...

- begin to suggest ways of defining the properties that I am testing (e.g., "strongest" structure, "runniest" liquid, "hardest" rock).
- use operational definitions to identify a variety of objects or living organisms.
- construct operational definitions when necessary and select appropriate ones for use in various experimental situations.

Summative Evaluation

LEVEL 1: At a SUPERIOR level, I always fully define the parameters of an experiment before I start, or, in the lower grades, I point out specific attributes that can be tested, and suggest the proper way to set up the test.

LEVEL 2: On the AVERAGE, I can rewrite a definition to ensure that it is testable, or, in the lower grades, I use words that define my observations or conclusions precisely; I look for definitions before beginning an experiment.

LEVEL 3: My work is SATISFACTORY if I am able to understand the need for specific definitions in doing experiments; I try to set up the parameters of my experiment before beginning; I recognize some of the words that may need definition.

LEVEL 4: My work would be UNSATISFACTORY if I am unable to define experimental attributes at all; I always act impulsively or randomly in experimenting; I consistently use the term "best" without analyzing any of the properties involved.

SO, IF MY TEACHER IS GOING TO EVALUATE MY ABILITY TO MAKE OPERATIONAL DEFINITIONS, THAT IS WHAT HE LOOKS FOR!

FORMULATING HYPOTHESES – WE MAKE EDUCATED GUESSES

FORMULATING HYPOTHESES IS THE PROCESS BY WHICH A NUMBER OF OBSERVATIONS ARE INCORPORATED INTO A GENERAL EXPLANATION.

Developing The Skill

MY TEACHER HELPS US LEARN THIS SKILL BY...

➥ keeping the hypotheses in line with our level of understanding.

e.g.,

K-1: If we dig down into soil, we will find rocks in the ground.

1-3: If we examine rock outcrops, we will find out that rocks are a part of the earth.

3-5: If we examine rocks from many different locations, we will find that there are different kinds of rocks.

6-7: If we examine a variety of rocks in several different ways, we will find that rocks are made up of different minerals.

➥ encouraging all of us to make hypotheses -
we share, list, trade and test them;
we try them out on our neighbors.

➥ helping us to carry out experiments
to support or refute our hypotheses.

➥ accepting our hypotheses
until we examine them through experimentation,
even if they are known to be false.

My teacher gives me ownership
of the hypotheses that I am testing!!!

Demonstrating the Skill

I DEMONSTRATE THIS ABILITY BY FOLLOWING A PROCESS...

1. **Ask a question**.
    a. Where do rocks come from?
2. **Make observations.**
    a. Collect rocks from many places.
    b. Examine rocky surfaces, rivers, etc.
3. **Evaluate the data.**
    a. Rocks come from many places on the surface of the earth.
4. **Provide a testable explanation.**
    a. Since rocks can be found just about anywhere I look, and since some hard parts of the earth, like mountainsides and riverbeds, are made of rock, I think that rocks are part of the surface of the earth.
5. **Call it a hypothesis.**
6. **Invite other people to test out my hypothesis and to try to prove me wrong.**

THAT'S SCIENTIFIC!

Evaluation of the Skill

IN GRADES 1-3, I SHOULD BE ABLE TO...

- give reasons for my scientific suggestions.
- test to prove my guesses.
- put several observations together and develop a way of explaining them all at once.

BY GRADES 4-7, I CAN...

- tell the difference between statements that are hypotheses and those that are not.
- decide which observations support a hypothesis or refute it.
- construct a hypothesis to explain the observations or data that I have collected.
- change my hypothesis if the data I collect are not consistent with it.

This is the very heart and soul of inquiry learning.

Children will learn what they discover for themselves, not what I tell them is true.

And hypotheses are at the root of concept learning.

When they have developed and tested their own hypotheses, they can truly understand the underlying concepts.

Summative Evaluation

LEVEL 1: At a SUPERIOR level, I am able to suggest general reasons for the observations and ways of testing my suggestions; I am able to separate statements that are not hypotheses from those which are, with reasons for my conclusions; I always carefully examine my experimental results to ensure a consistent relationship with each hypothesis; I am willing and able to make changes to my hypotheses if necessary.

LEVEL 2: On the AVERAGE, I can state a hypothesis when given several observations and possible ways of correlating the data; I can decide whether or not new experimental results are consistent with the hypothesis.

LEVEL 3: My work is SATISFACTORY if I am able to identify a workable hypothesis for a set of observations following a detailed examination of the data and a discussion of the underlying concepts.

LEVEL 4: My work would be UNSATISFACTORY if I am unable to even choose a satisfactory hypothesis from several possible choices after examining the observations and the concepts; if I rely solely on guesses to explain any observations.

SO, IF MY TEACHER IS GOING TO EVALUATE MY ABILITY TO MAKE HYPOTHESES, THAT IS WHAT SHE LOOKS FOR!

Designing Experiments – We Organize Our Plans

EXPERIMENTATION IS THE PROCESS BY WHICH WE INVESTIGATE A PROBLEM.

The Process...

CHOOSE a researchable question.

RECOGNIZE the problem being investigated.

PLAN an experiment to obtain information that will answer the question.

IDENTIFY one or more of the variables involved.

MANIPULATE the variables so that the effect of one identified variable can be observed and measured.

COLLECT data.

INTERPRET and **ORGANIZE** the data.

COMMUNICATE the results of the investigation.

Developing the Skill

➥ My teacher realizes that experiments are often carried out in a haphazard manner at first.

➥ Thus, I have assistance in developing more systematic approaches to solving my problem.

➥ I am helped as I try to integrate all the steps in the process of experimentation.

➥ I am shown how to recognize valid results.

➥ I am taught to integrate all the other science process skills in my search for a solution.

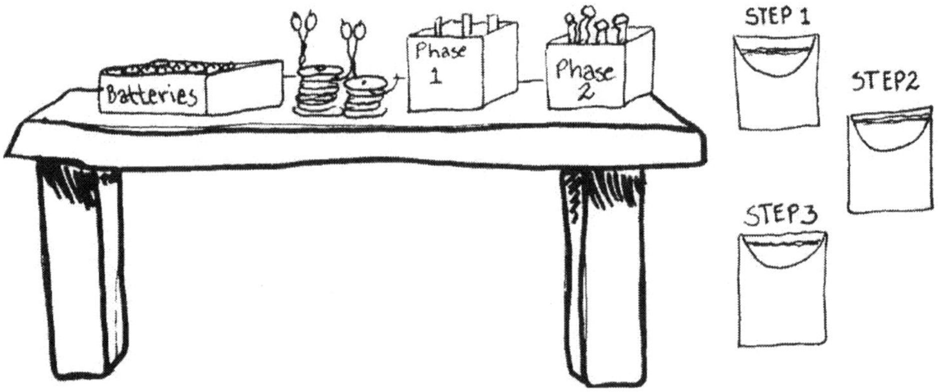

Evaluating the Skill

IN GRADES 1-3, I SHOULD BE ABLE TO...

- realize that there are many ways of finding an answer to a question.
- devise a way of finding an answer to a problem.
- organize my activities when trying to solve a problem.

BY GRADES 4-7, I CAN...

- see a systematic approach as I work to find the answer to a given question.
- follow the appropriate procedure as I work toward the solution of a problem.
- construct hypotheses and investigate them in a logical manner.

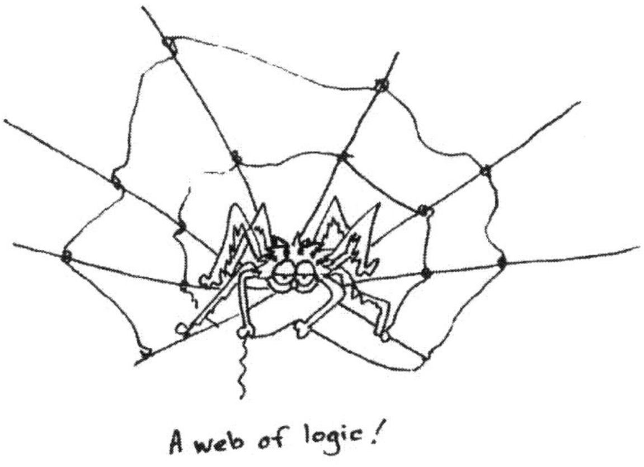

Summative Evaluation

LEVEL 1: At a SUPERIOR level, I approach all problems systematically and consistently set up experimental procedures to study a problem before I make conclusions.

LEVEL 2: On the AVERAGE, I can set up an experiment to test specific problems using logical experimental procedures.

LEVEL 3: My work is SATISFACTORY if I am able to proceed through the steps of an experiment as presented and know the basic steps and skills required to complete experiments satisfactorily.

LEVEL 4: My work would be UNSATISFACTORY if I fail to follow even the most basic of sequences when conducting an experimental investigation.

SO, IF MY TEACHER IS GOING TO EVALUATE MY ABILITY TO DEVELOP EXERIMENTS, THAT IS WHAT HE LOOKS FOR!

A SUMMARY OF THE SKILLS

THE BASIC PROCESS SKILLS

OBSERVING	What can I notice?
INFERRING	How can I explain that?
COMMUNICATING	Let me tell you what I've done.
MEASURING	Which one is greater?
CLASSIFYING	What is the pattern in this?
SPACE/TIME RELATIONSHIPS	In what order did these things happen? At what rate did that change?

THE ADVANCED PROCESS SKILLS

PREDICTING	I think the results will be...
CONTROLLING VARIABLES	What things are going to influence this test?
INTERPRETING DATA	What does all this information tell me?
OPEERATIONAL DEFINITIONS	Let me make it clear what I am really testing.
FORMULATING HYPOTHESES	I think that all of this can be explained by this simple statement.
DESIGNING EXPERIMENTS	How can I test this to see what else I can find out?

LET'S SET THE STAGE FOR THE STUDY OF CONTENT!

SECTION 6 - LOOKING AT CONTENT

ANIMALS

MATTER

PLANTS

ENERGY

EARTH

HEALTH

SPACE

TECHNOLOGY

CONTENT - TO BUILD ON

Animals – We Appreciate Living Things

Studying a Variety of Animals

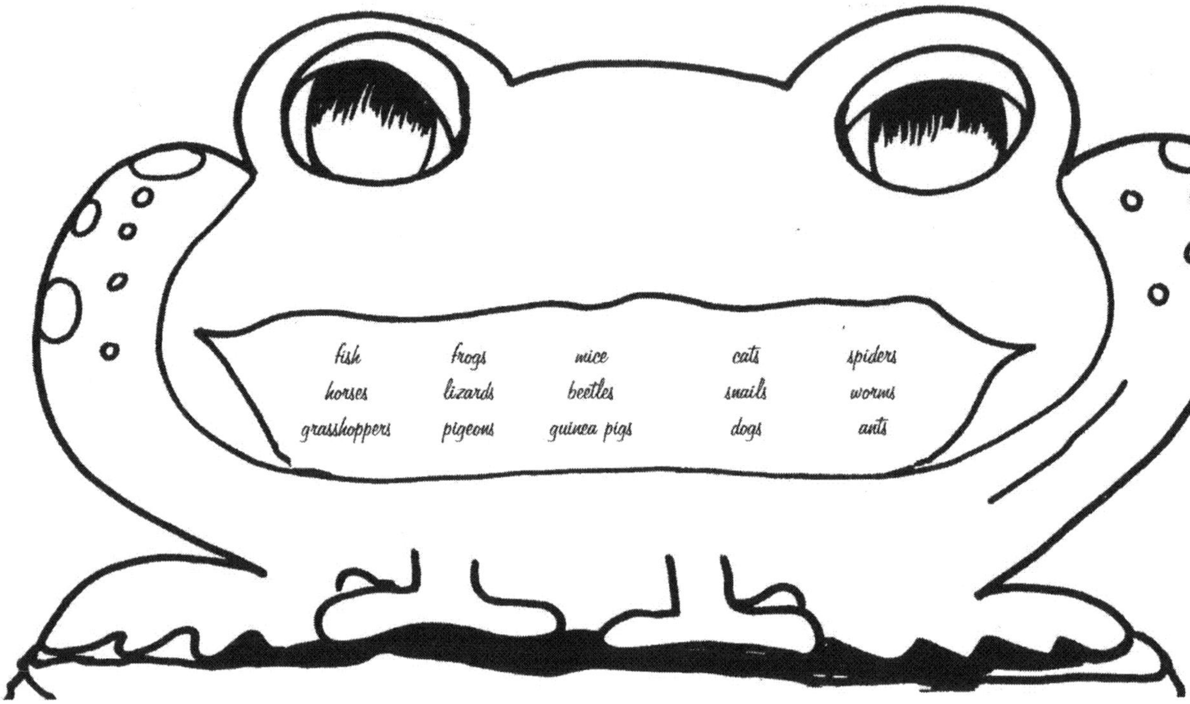

Exploring Habitats

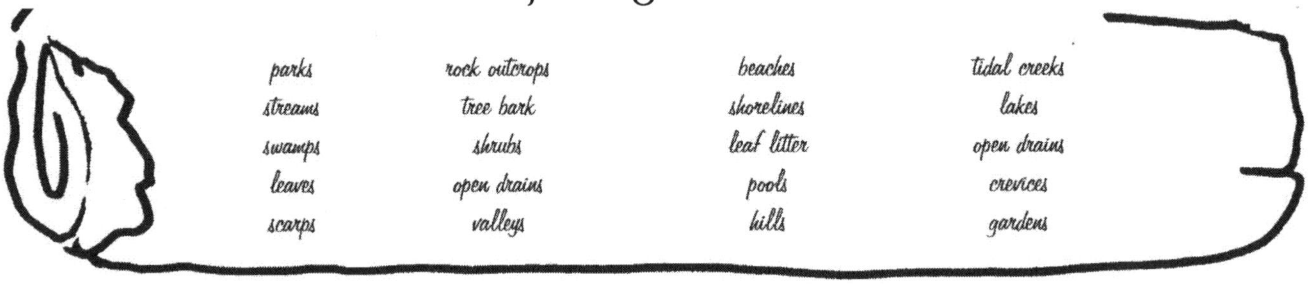

THERE ARE SO MANY THINGS ABOUT ANIMALS THAT WE CAN INVESTIGATE...

➤ physical characteristics -
body parts, legs, eyes, coverings, etc.

➤ movement:
method, speed, length of pace.

➤ preferred food and drink.

➤ habitat:
terrestrial/aquatic, dry/moist,
sand/rock/bark/tree.

➤ behavior:
response to noise, light, heat, etc.

➤ life cycle:
eggs, new born, larval (pupal) stages,
dependence on parents.

PLANTS – WE EXAMINE OUR ENVIRONMENT

I'LL DEVELOP A BETTER UNDERSTANDING OF THE PLANT WORLD IF MY TEACHER...

➤ gives me a chance to grow and to handle plants of my own in the classroom.

➤ gives me the opportunity to observe many different kinds of plants in their natural setting.

➤ selects plants that represent the great diversity that is to be seen in the plant world.

THERE'S A WORLD OF LIVING THINGS OUT THERE FOR US TO LOOK AT...

fungi
vegetables
trees
algae
lichen

mosses
fruit
ferns
grasses

THERE ARE MANY ASPECTS OF PLANT LIFE THAT I'D LIKE TO INVESTIGATE MORE CLOSELY.

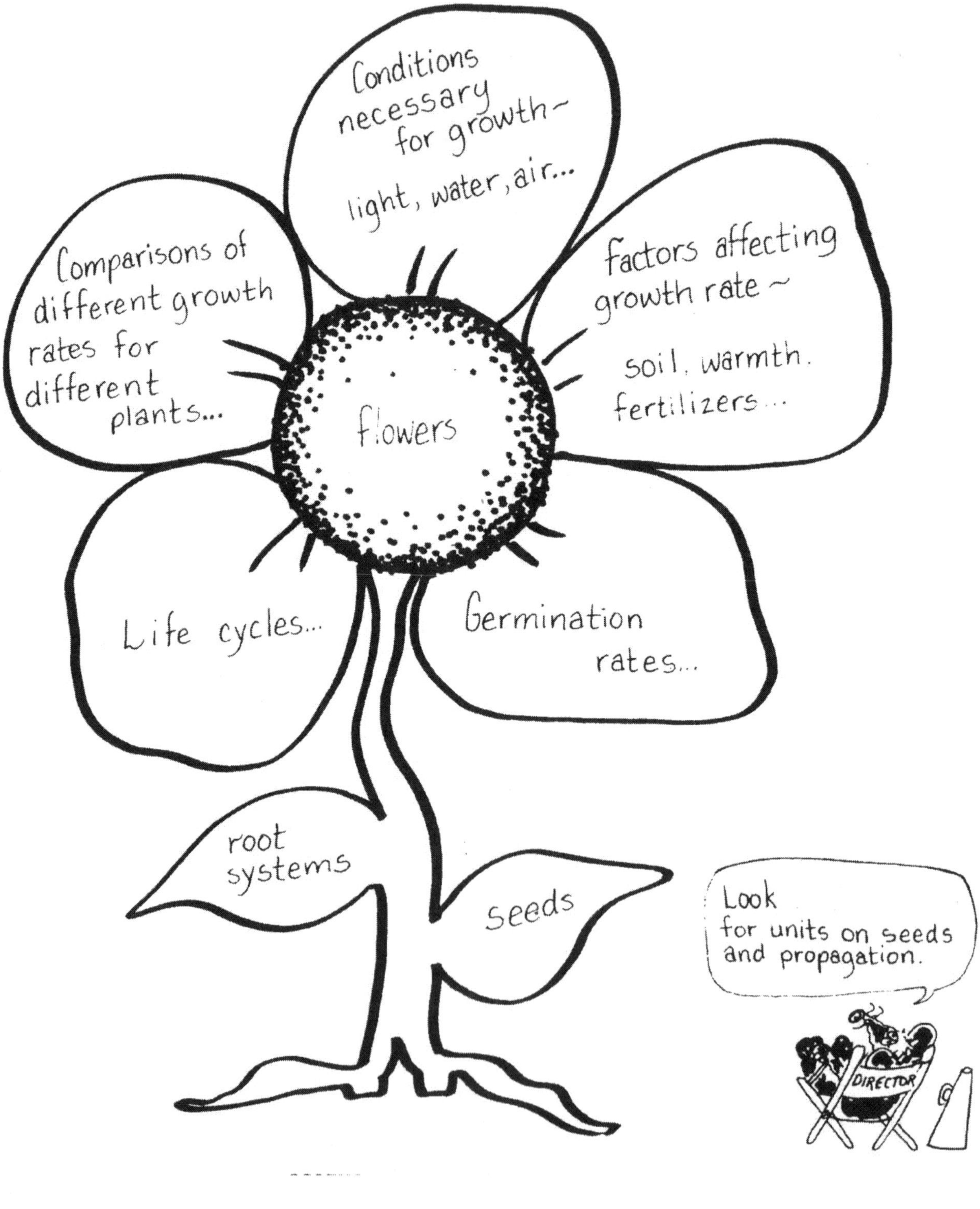

LOOK AT ALL THE ACTIVITIES THAT WE CAN DO OUTSIDE THE CLASSROOM.

➤ **mapping** where different plants grow.

➤ **comparing / contrasting** habitats.

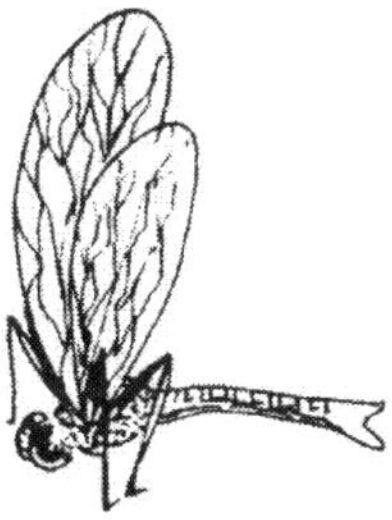

➤ **studying** the many animals that live on plants.

➤ **classifying** using leaves, bark, size, color, branching systems, habitat, etc.

➤ **observing diversity** in structure, usefulness, hardiness, need for shade, etc

➤ **observing** the effect of changing habitats.

AND LET'S EXPLORE MANY HABITATS.

swamps

forests

river beds

tidal pools

gardens

yards

beaches

(Look back at the list of animal habitats— we'll find lots of plants there, too!)

HEALTH - WE LEARN ABOUT OURSELVES

HEALTH SCIENCES

INVOLVE EXAMINING AND UNDERSTANDING MANY ASPECTS OF THE HUMAN BODY.

Germs and Cleanliness

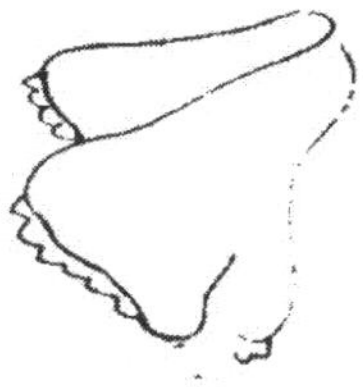

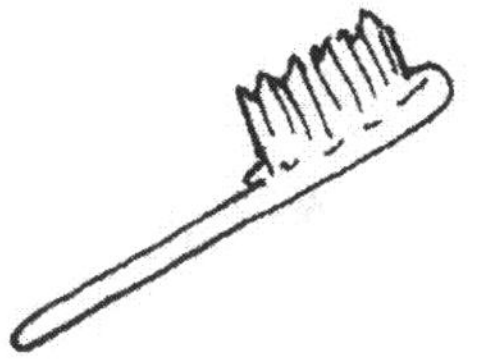

- bacteria and antiseptics
- viruses and diseases
- immunization
- the body's defenses
  (mucus, cilia)
- (white blood cells, antibodies)
- allergies

Human Physiology

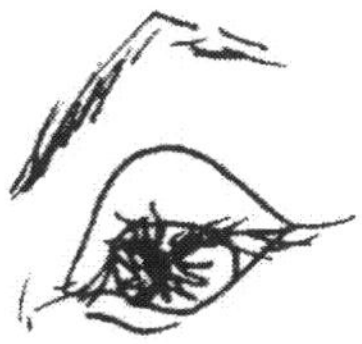

- systems, organs, tissues, cells
- anatomy
- physical growth and maturation
- posture, balance, and strain
- muscle power
- the senses
- protecting our sense organs

- heredity and genetics
- chromosomes
- DNA and genes

Nutrition

- food intake
- vitamins and minerals
- keeping food safe
- sanitation and pollution
- drinking water
- dental hygiene

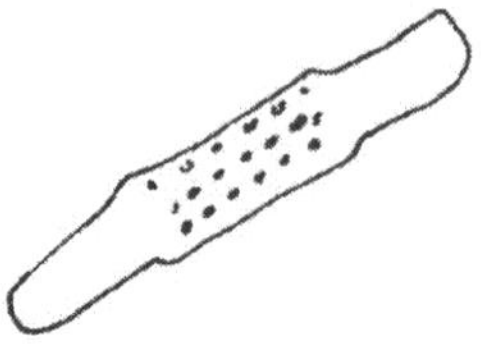

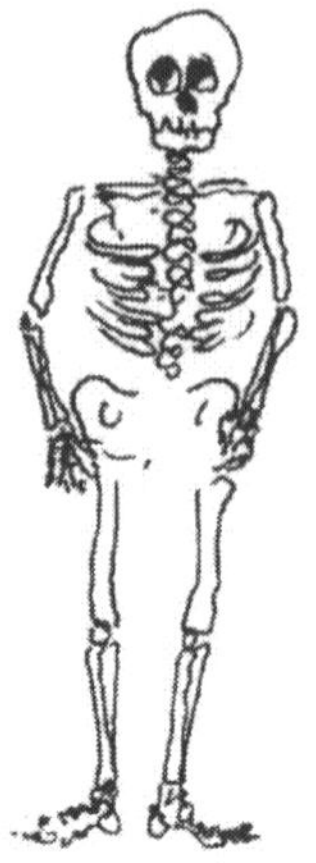

Social Interaction

- at school, at home, as a guest, in the community
- friendships
- differences between people
- community health and safety services
- hospitals, food inspectors, fire inspectors

Safety

- clothing for protection, appearance and safety
- home, road, bicycle, and water safety

Emotions

- joy, excitement, fear, anger, anxiety
- emotional growth
- moods and behavior
- sleep, fatigue and rest
- exercise and fitness
- leisure time

Learning Resources

As well as reading textbooks and doing research reports,
we talk to people in our community that are concerned with health and safety.

We examine ourselves and others through experimentation and questioning.

We study food and food producers.

We join young people's groups and clubs,
so that we can learn to get along with others
in many different settings and circumstances.

MATTER – WE EXAMINE MANY THINGS

MATTER
IS THE NAME THAT IS GIVEN
TO ALL THE NATURAL AND MAN-MADE SUBSTANCES
WHICH MAKE UP THE UNIVERSE.

WE EXAMINE A LOT OF NATURAL AND MAN-MADE SUBSTANCES...

sugar	*kerosene*	*fruit*	*feathers*	*paint*
vinegar	*starch*	*staples*	*butter*	*pencils*
milk	*cement*	*wire*	*rubber*	*soil*
balls	*cardboard*	*wood*	*plasticene*	*putty*
matches	*glass*	*plastic*	*chalk*	*fabrics*
charcoal	*soap*	*rocks*	*detergent*	*coins*
string	*flour*	*disinfectant*	*bones*	*perfume*
powder	*candles*	*metals*	*water*	*paper*
chewing gum	*fishing line*			

AND STUDY A WIDE VARIETY OF PROPERTIES...

buoyancy · *size* · *shape* · *color* · *elasticity*

stability · *strength* · *odor* · *transparency* · *temperature* · *volume* · *noise production* · *heat and electrical conductivity* · *effects of heat, solvents, abrasion*

texture · *force* · *weight* · *speed* · *taste*

USING A VARIETY OF STANDARD AND CLASS-MADE INSTRUMENTS.

rulers
hand lenses
balances

tapes
clocks
egg timers

pendulums
shadow sticks
thermometers

ENERGY – WE UNDERSTAND ENERGY SOURCES

Forms and Sources of Energy

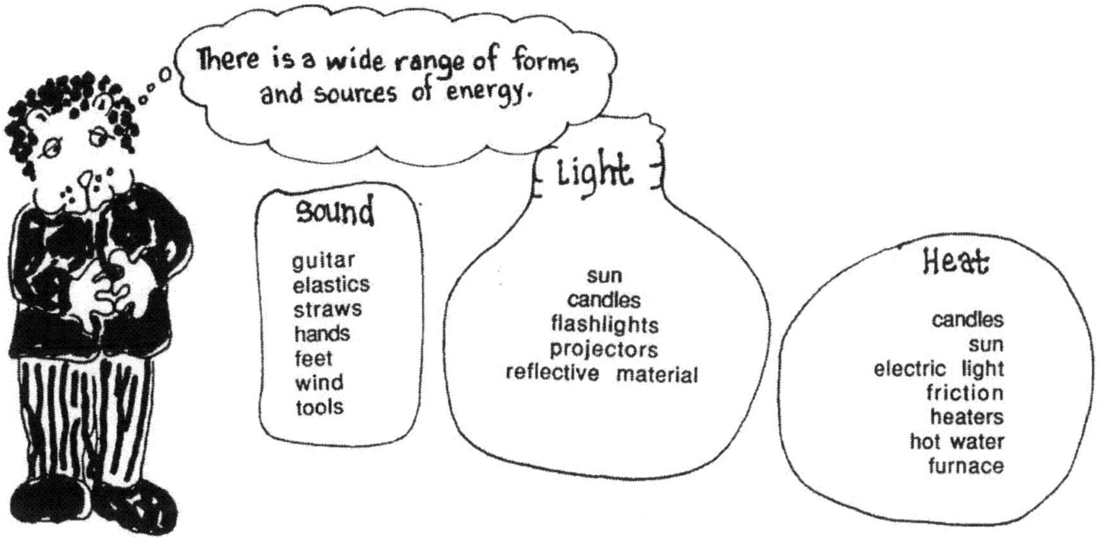

Energy of Motion

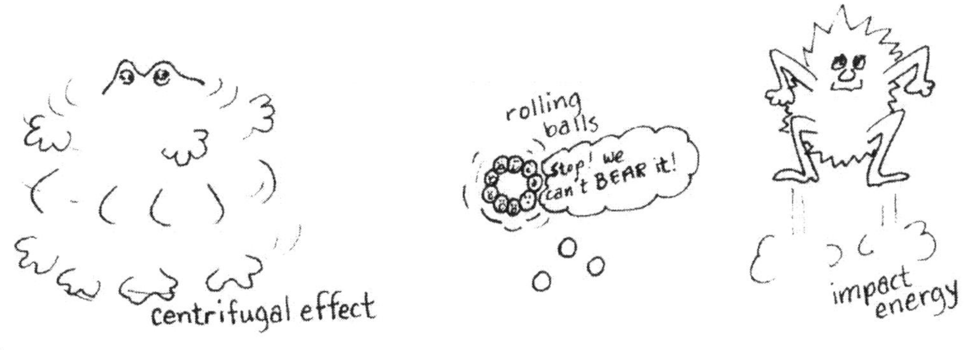

Investigating Energy

There's a lot of potential For the study of energy!

➤ where it can be seen
➤ how it is used
➤ size of force it generates
➤ how it can be controlled
➤ whether it can be transmitted, absorbed or reflected
➤ what effect it has on materials

THERE ARE MANY WAYS THAT WE CAN INVESTIGATE ENERGY...

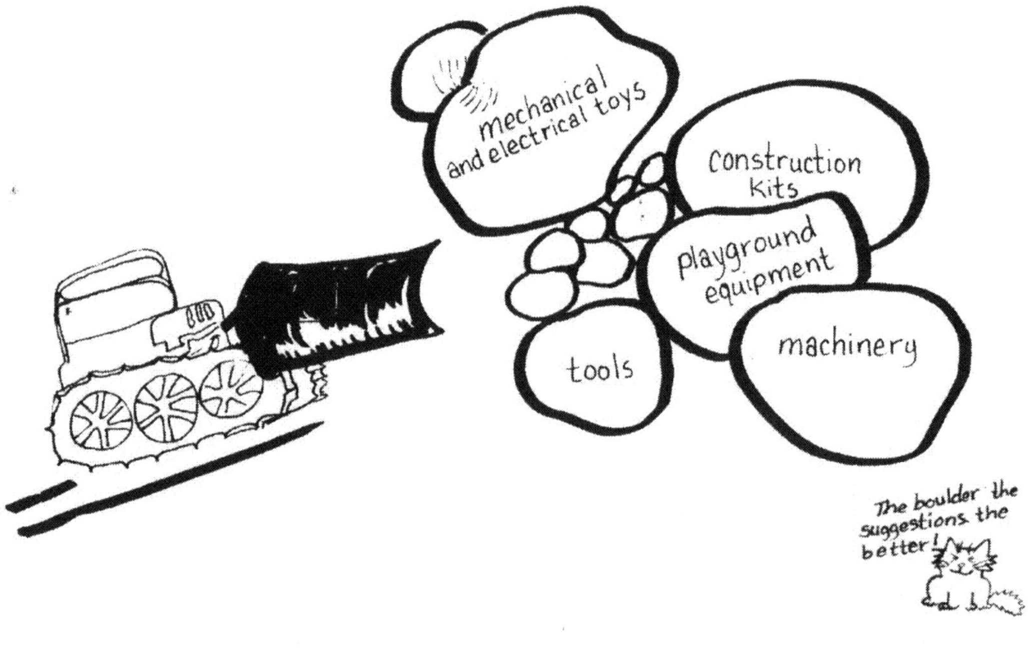

EARTH – WE UNDERSTAND OUR OWN PLANET

THE STUDY OF THE EARTH INCORPORATES THE STUDY OF ANIMALS, PLANTS, MATTER, AND ENERGY.

IN ORDER FOR ME TO DEVELOP AN AWARENESS OF THE GREAT DIVERSITY OF EARTH STUDIES...

I need to learn about the interrelationship of earth, environment, and habitat.

In order to survive, an organism must have an environment that will satisfy its many needs.

This environment can be on, under or above the earth's surface.

The geological formations, the climate, and the presence or absence of other life forms can determine what form of organisms may exist in any particular region.

rock outcrops beaches tidal creeks streams shorelines
lakes swamps crevices scarps valleys hills

Environments are constantly changing.
Plants grow, animals come and go,
rivers change their courses, mountains erode, climate changes.
Changes in one environment can affect neighboring environments,
as well as those at a great distance.

No creature changes its environment as much as man does.
However, we still are not able to fully predict the effects of these changes,
or the effects of attempts at conservation.

There are many differences between individuals of different species,
and this helps them to adapt to different environments.

Environments are classified on the basis of their vegetation and climate.

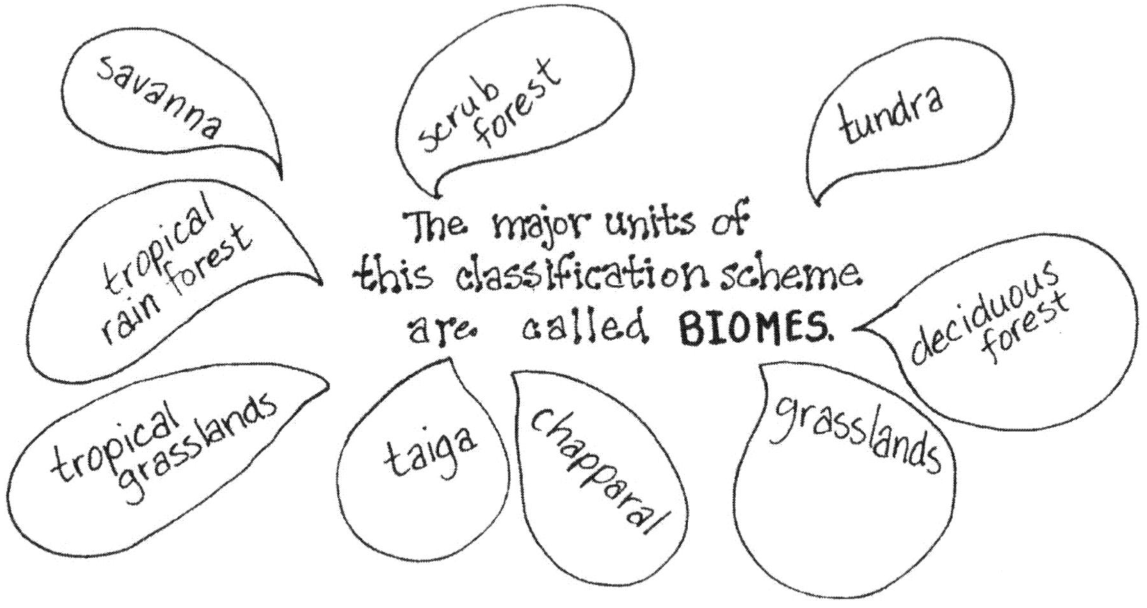

Environments and biomes merge with one another
Any two environments located in the same biome can be said to be related.
Plants and animals may range across two or three biomes.

THERE ARE SO MANY THINGS WE CAN INVESTIGATE ABOUT THE EARTH'S SURFACE...

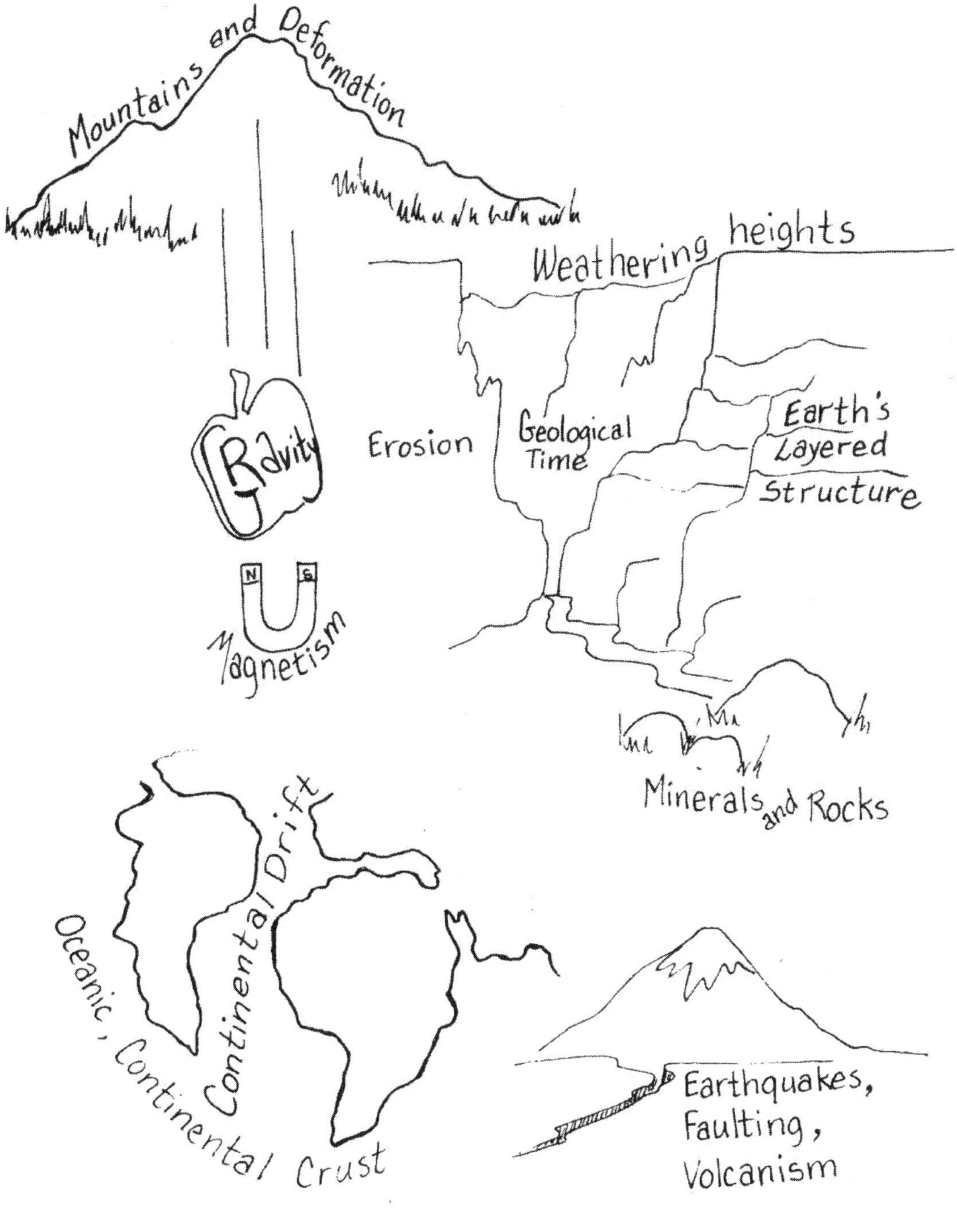

IN STUDYING THE ATMOSPHERE, WE CAN LEARN ABOUT MUCH MORE THAN JUST THE WEATHER...

Atmospheric Regions

IONOSPHERE: (60-1000 km)

- exosphere
- thermosphere
- mesosphere

- stratosphere
- troposphere

Pressure, Temperature, Humidity

Precipitation

- clouds
- fog
- dew
- rain

Winds and Wind Systems

- hurricanes
- tornados
- typhoons
- thunderstorms

Climatic Regions

- polar
- temperate
- tropical

SPACE –
WE STUDY BEYOND THE EARTH

THERE'S AN INFINITY OF THINGS THAT I CAN STUDY AS I TRY TO LEARN ABOUT OUTER SPACE...

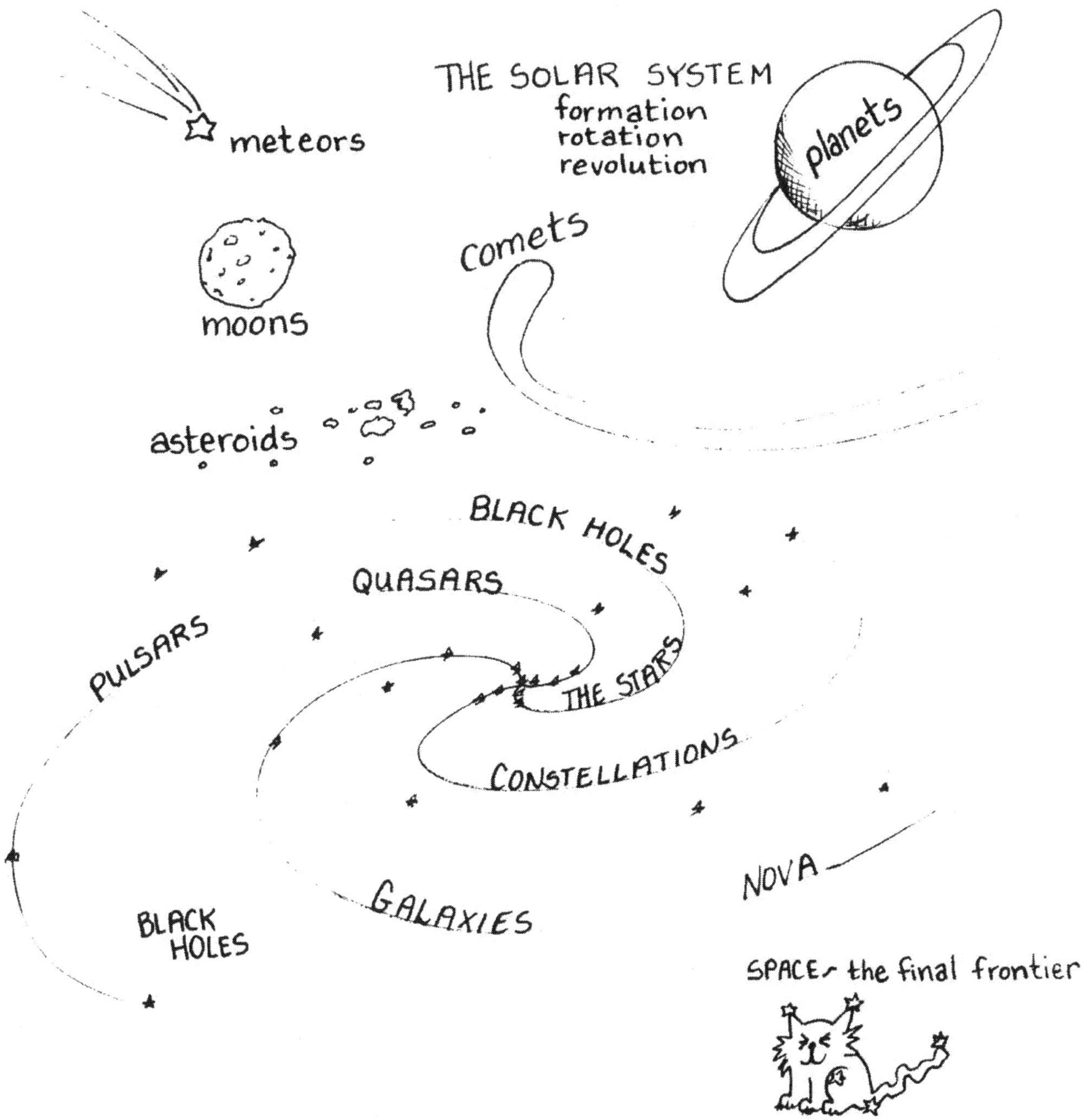

The Solar System

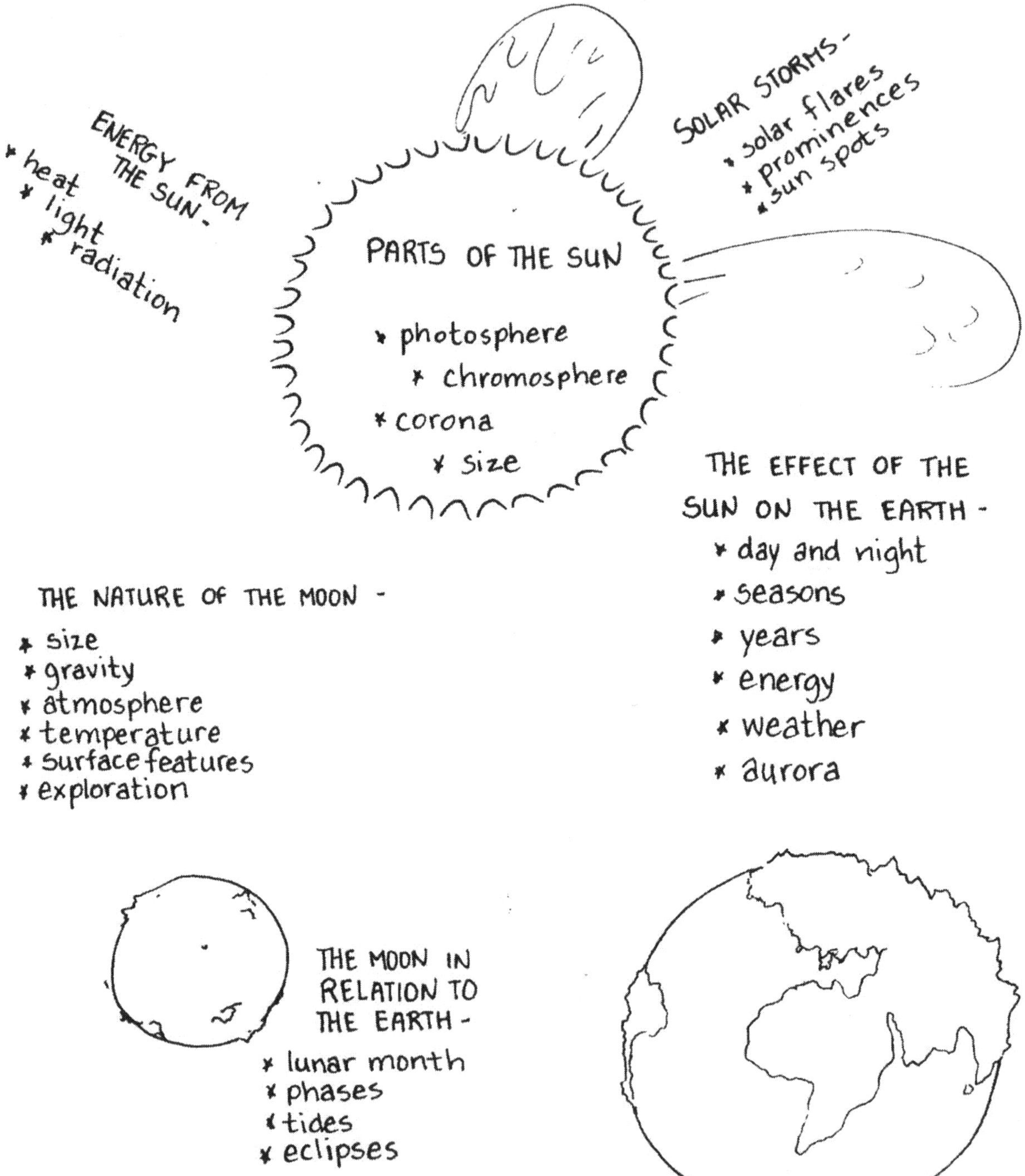

AND THERE IS MUCH I'D LIKE TO LEARN ABOUT SPACE TRAVEL...

MAN IN SPACE

- The history of space travel
- Exploration of the planets
- American, Russian, French, and Canadian and other space programs
- Manned space travel
- Problems for astronauts in space
    - Acceleration, weightlessness
    - Air, heat and cold
    - Radiation, space sickness
    - Physical changes
    - Landing
- Flight to the moon
- The International Space Station
- Exploration of Venus and Mars

ROCKETS, SATELLITES, AND SPACECRAFT

- Orbiting and escaping the earth
- Staying in orbit
- The path of a satellite
- Tracking satellites and spacecraft

TECHNOLOGY ~ WE EXAMINE THE BEST USES OF SCIENCE

TECHNOLOGY IS THE PRACTICAL APPLICATION OF SCIENTIFIC RESEARCH.

The Value of Studying Technology

It extends our ability to change
what nature has created,
and it also directly alters the way we live.

Every aspect of our lives is affected by
the benefits and losses associated with the growth of science and technology.

Technology also allows us to conduct the research
that helps us to understand nature,
and the processes of scientific discovery.

In order to fully appreciate science,
we must consider not only the results of pure science,
but also the impact of scientific discoveries
on the way we live and the way our societies grow.

Areas of Study

TECHNOLOGY IS AN AREA THAT I'D LIKE TO PLUG INTO.

medicine, hospitals, health care

recreation, sports, equipment

toys, games, amusement park rides

houses, appliances, heating

skyscrapers, bridges, pyramids

video games, calculators, computers

tape recorders, record players, compact disc players

the forest industry

mining technology

telecommunications (e.g., telephones, radio, T.V., computers, etc.)

transportation (e.g., trains, cars, boats, planes, etc.)

experimental techniques

garbage (e.g., problems of disposal, sewage systems, etc.)

food production and distribution

AND THERE'S A MULTITUDE OF WAYS THAT I CAN STUDY DIFFERENT TECHNOLOGIES.

Research	Books, magazines, DVDs, interviews, T.V. programs, field trips, internet.
Examination	Looking at different machines and tools to discover how they work.
Experimentation	Looking at a need and devising ways of meeting the need.
Imagination	Drawing and designing machines that can do many different things.
Usage	Using many different machines and devices, while paying attention to how they work and what they are used for.

Solar powered
wind-up
battery-powered
electric
wind-powered

My intention in all of this is to gain a greater understanding of the technology that exists in my world, why it has been developed, and how it makes my life easier, safer, more productive or more enjoyable.

SECTION 7 -
THE EXPERIENTIAL LEARNING CYCLE

EXPERIENCING

SHARING

INTERPRETING

GENERALIZING

APPLYING

PROCESSING

EXPERIENTIAL LEARNING IS...

- Initial limited orientation followed by participation activities in a real setting.
- Examination and analysis of experiences so one can identify the effect of one's actions.
- Opportunities to generalize and summarize.
- Opportunities to try out conclusions.

THEORY: Demonstration > Practice > Feedback > Application
EXPERIENCE: Sharing > Interpreting > Generalizing > Applying

Experiencing

- Generate individual data from one or more of sensing, thinking, feeling, wanting or doing modes. (This moves the group to the next stage, or more deeply in to this one.)

What is going on here? What else?
How do you feel about that? And?
What do you need to know to...?
Would you be willing to try...?
Can you be more specific?
Could you offer a suggestion?
What would you prefer?
What are your suspicions?
What is your objection?
Can you say that in another way?
Can you say more?
What is the worst/best thing that could happen?
If you could guess at the answer what would it be?

Sharing

- Report the data generated from experience.

Who would like to volunteer to share?
Who else?
How did you feel about that?
Who else had the same experience?
Who reacted differently?
Were there any surprises? Puzzlements?
How many felt the same?
How many felt differently?
What did you observe?
What were you aware of?

Interpreting

➤ Make sense of the data for the individual and the group.

How did you account for that?
What does that mean to you?
How was that significant?
Was that good/bad?
What struck you about that?
How do those fit together?
How might it have been different?
Do you see something operating here?
What does this suggest to you about yourself/your group?
What do you understand better about yourself/your group?

Generalizing

➤ Develop testable hypotheses and abstractions.
(This leads toward abstracting from specific knowledge.)

What might we draw/pull from that?
Is that plugging into anything?
What did you learn/relearn?
What does that suggest to you about ----- in general?
Does that remind you of anything?
What principle/law do you see operating?
What does that help explain?
How does this relate to other experiences?
What do you associate with that?
So what?

Applying

➤ Bridge the present and the future by understanding and/or planning how these generalizations can be tested in a new place.
(Utilizes the learning in the real world.)

How could you apply/transfer that?
What would you like to do with that?
How could you repeat this?
What could you do to hold onto that?
What are the options?
What might you do to help/hinder yourself?
How could you make it better?
What would be the consequences of doing/not doing that?
What modifications can you make work for you?
What could you imagine/fantasize about that?

Processing

➤ Examine the entire experience as a learning experience through the solicitation of feedback.

How was this for you?
What were the plusses/minuses?
How might it have been more meaningful?
What is the good news/bad news?
What changes could you make?
What would you continue?
What are the costs/benefits?
If you had it to do over again, would you?
What additions/deletions would help?
Any suggestions?

10 RULES FOR ELEMENTARY SCIENCE TEACHERS

1. Allow every student a chance to take part in the experiments.

2. Make the sessions non-threatening.

3. Be patient with the children.

4. Allow the children to control the amount of time spent on the experiment.

5. Use open-ended questions.

6. Give children ample time to answer questions.

7. Don't expect "standard" answers or reactions from children.

8. Accept divergent answers.

9. Be sure to encourage observation.

10. Look for interesting ways to extend an activity.

SECTION 8A - ACTIVITIES FOR GRADES K-3

CLASSROOM ANIMALS

LOOKING AT SEEDS

OURSELVES

LOOKING AT AN APPLE

LIGHT

THE LIVING EARTH

FLIGHT

PENDULUMS

CLASSROOM ANIMALS

ACTIVITIES FOR THE STUDY OF ANIMALS (K-3)

UNIT OBJECTIVES

Skills

Observation

* Study a variety of animals and pets in a controlled environment.

(While all the senses should be used in observation, their contribution varies with the nature of the object being investigated.)

Inferring

* Suggest reasons for various behavior patterns.
* Make inferences to fit observations as to how animals obtain water, air, shelter and protection.
* Develop divergent thinking skills by suggesting several inferences to explain an observation or event.

Communication

* Describe the variety of animals and habitats studied.
* Draw pictures, describe observations, share results.

Attitudes

Curiosity

* Examine a variety of pets and classroom animals.

Caring

* Handle small animals with care.

Concepts

Change

* Consider the changes in the animals over the year, in different seasons, at home or at school.

Interaction

* How we handle a pet has a lot to do with how it will behave.
* Pets cause changes in households.

REMEMBER, DO NOT DO ANYTHING TO HARM YOUR ANIMAL!

Unit Activities

Consider the Animal's Characteristics

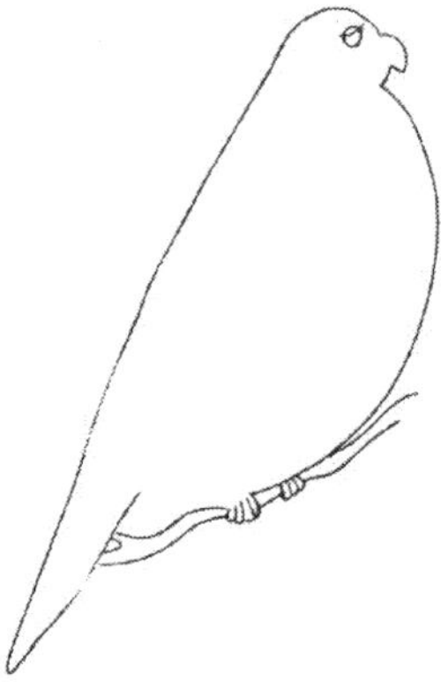

Examine many things...

height, length, girth, weight
color, movement

ears, eyes, feet, toes, hands, fingers, teeth

body covering
growth rate
pulse rate

Consider the Animal's Reactions

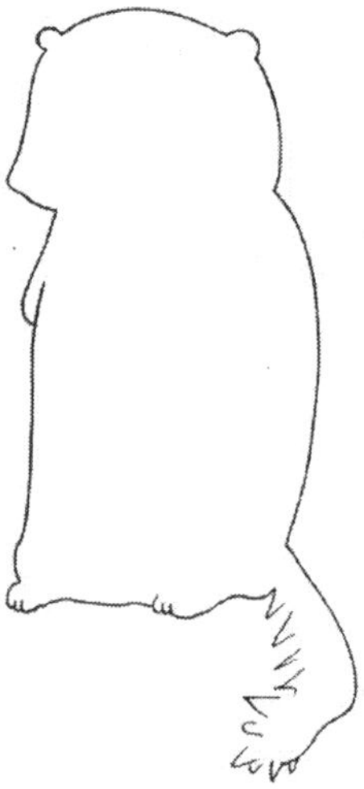

Does the animal react to...

light? heat?
dampness? touch?
smell? sound?
color?
people?
other animals?
its parents?

When are the reactions positive? (A motion toward the stimulus)
When are the reactions negative? (A motion away from the stimulus)

Consider Your Animal's Babies

Are the babies born alive or hatched?
Are there multiple births?
Are the young identical?
Are they born blind?
Do the young nurse?
Are the babies like the parents?

Consider the Animal's Daily Life

Does your animal sleep most of the day?
Does it play a lot?
Does the animal fight a lot?
Does it like people or not?
What kind of a noise does it make?
Will it hide from you?
Does it have a best friend?
Can it protect itself well? How does it do this?
Does your animal clean itself?

Consider the Animal's Eating Habits

Does the animal prefer...

vegetables?
grasses?
meats?
nuts, grains, seeds?
milk?
water?
suckling?

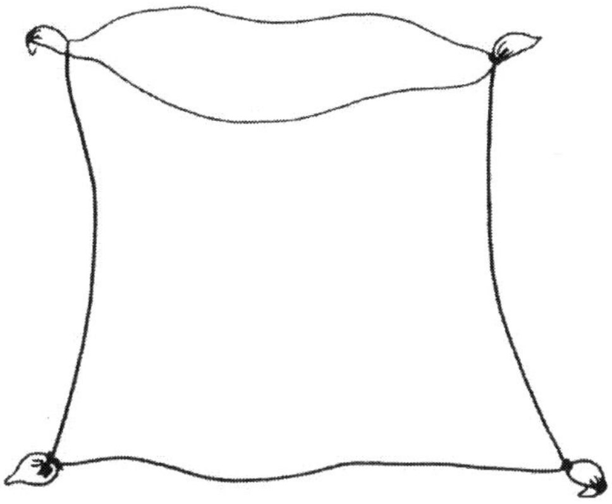

How much food does your animal consume each day?

Consider Lots of Other Things

Look at your animal's natural habitat.

Compare it with its close relatives.

Look for special adaptations.

What is its mating behavior?

What is its learning ability?

What is its role in society?
(e.g., pet, beast of burden)

Try a Few More Activities

* Build cages.

* Make clay models.

* Make animal masks.

* Reconstruct skeletons.

* Find animal symbols that are used by companies.

* Look at animals on stamps.

* Imitate animal movement.

* Study animal tracks.

* Study animals in circuses.

* Look at songs and poems about animals.

* Consider animal rights.

LOOKING AT SEEDS

ACTIVITIES FOR THE STUDY OF PLANTS (K-3)

UNIT OBJECTIVES

Skills

Measuring

* Chart the growth of different plants.

Classifying

* Classify seeds according to a variety of characteristics.

Space and Time Relations

* Explain the sequence of development of growth in a plant.

Attitudes

Persistence

* Carry through the study of a plant from preparation of the seeds through planting and growth.

Inventiveness

* Consider a variety of locations for finding seeds.

* Prepare a seed decoration.

Concepts

Interdependence

* Plants respond to stimuli from the physical and living environment.

Variety/Commonality

* Plants grow and change at different rates.

* When seeds are planted, the new plant has the same characteristics as the old one.

* Plants of the same species may vary in size and appearance under different environmental conditions.

UNIT ACTIVITIES

Look for Seeds Growing on Plants

Where are the seeds?
(E.g., at the end of a branch, on a long stem,
on the branches, where the flower was.)

Is there a seed case?
What shape is the seed/seed case? What color is it?

How are ripe seeds different from "green" or immature seeds?

What seeds are covered with soft flesh? (such as a peach)

Find seeds that are fluffy and light. ("fly away" seeds)

Which plants have seeds that are prickly?

Find plants that have seeds in pods.

Collect and Classify Seeds

CONSIDER TEXTURE.

Which seed cases feel hard?
Find some soft seeds, such as a tomato seed.
Find some hard seeds that grow inside a soft seed case (such as grapes).

CONSIDER SIZE.

Do all large plants produce large seeds?
Are most small seeds the same shape?
Group seeds into very big, middle sized, and small.

CONSIDER SHAPE.

Which plants have flat seeds?
How are round seeds different from one another?
Discuss different shaped seeds that have been collected.

CLASSIFY SEEDS IN OTHER WAYS.

Sort seeds according to color.
Sort seeds according to smell.
Sort seeds into those that are eaten... by people, by animals.
Discuss seeds that are not safe to eat.

Soak Seeds

SOAK TWO LARGE SEEDS IN WATER. COMPARE THEM REGULARLY.

What changes were first noticed?
When were they first noticed?
Describe the changes in texture, color, shape, etc.

Compare a large seed that has been soaking for several days with a dry seed of the same type.
How does the skin feel? Does it peel easily?
What does the seed look like without the skin?
How many parts does it have?

Split each seed in half. What do you notice?

Grow a Seed

PUT SOME SEEDS IN SOIL IN AN EGG SHELL OR ON COTTON WOOL IN A SHALLOW CONTAINER.

What does a seed need to make it grow?
What is the first change to be seen in the seed?
Can you see when the seed is beginning to grow?
Which part of the plant appears from the seed first?
When does the first leaf appear?
What happens to the seed as the plant grows?

RECORD THE GROWTH OF SIMILAR PLANTS.

Examine the plant on a regular schedule (e.g., every 3 days).
As the plant grows, cut strips of paper to the same height as the plant.
Glue them to a chart to record growth.
Does the plant seem to grow regularly or irregularly?
Do the leaves appear at the same height?
Do the leaves get higher as the plant grows?

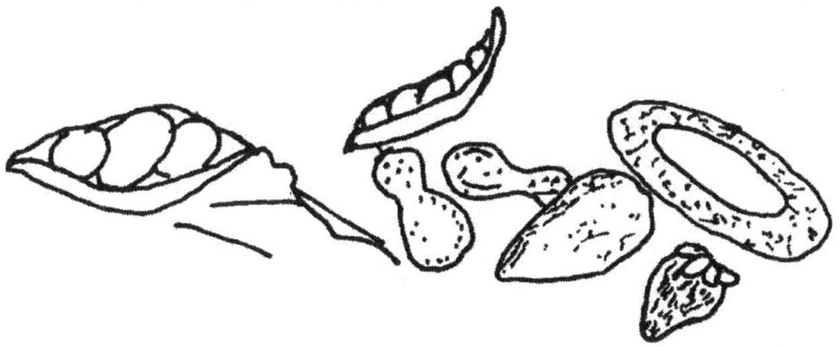

Examine Seed Usage

EXAMINE SEEDS THAT ARE USED BY PEOPLE...

...for food, in gardens, as food for animals, for decorations.
Sort the seeds according to which make the best decorations (for a necklace, wall plaque, dried arrangement). Consider color, size, and shape.

EXAMINE EDIBLE SEEDS.

What types of seeds may be eaten raw?
What are some different ways that seeds are prepared for cooking?
What sorts of seeds are crushed to make cooking oils?

NOTES FOR THE TEACHER

"Fluffy" seeds: thistle, dandelion, etc.
Prickly seeds: clover, grass, cotton, etc.
Pod seeds: sweet peas, cassia, etc.
Flat seeds: passion fruit, cucumber, pumpkin, etc.
Big seeds: bean, coconut, mango, etc.
Small seeds: carrot, apple, orange, wheat, marigold, etc.

The COTYLEDON is the food store of a seed. The seeds of wheat, oats, barley, rice, and corn are monocotyledon, while seeds of apples, oranges, marigolds, and radish are dicotyledon. This is readily seen when the seeds are soaked, as the dicotyledons split into two parts.

OURSELVES

ACTIVITIES FOR THE STUDY OF HEALTH (K-3)

UNIT OBJECTIVES

Skills

Predicting

* Make predictions before doing activities.

Controlling Variables

* Examine only one item at a time.

* Chart the relationships when changing variables.

Interpreting Data

* Prepare graphs and charts.

* Interpret graphs and charts.

Attitudes

Caring

* Work carefully with partners.

* Be aware of the body's needs.

Critical Thinking

* Accept the SUPPORTED findings of your classmates.

* Be sure that information is recorded accurately.

Concepts

Continuity / Conservation

* Our bodies are constantly changing,
But much of us stays the same as we grow and change.

Unit Activities

How Tall Are You? How Heavy Are You?

All members of the group should measure each other's height.

Who is the tallest?
Who is the shortest?
Who has the longest legs?
Record your results.

Do the same while you are sitting on a chair.
Is there much difference?
Record your results.

Weigh everyone in the group.
Record your results.
Compare the relationship between height and weight.

Hop, Skip and Jump

Each person does a hop, skip, and jump from a starting point.
Measure the total distance.
Does the person's height make any difference to how far they can jump?

What Area Does Your Silhouette Cover?

Ask a friend to lie down on a squared paper.
Draw all around him/her.
Count the squares (those that are greater than half size).
What is the area?

Shapes of Faces

Look at the faces of some of your friends.
Make a list of the ways in which the faces differ.
Make a drawing of all the different face shapes you find.
(e.g., round, oval, heart shaped, square, oblong, etc.)

Measuring Around Heads and Chests

Measure around each other's heads—just above the eyebrows.
Measure some older and some younger children.
Compare the results.

Do the same with your chest.

Breathe out, empty your lungs, and measure your chest.
Breathe in as much as you can, and measure again.
How much did your chest expand?

Walk in a Straight Line

Walk in a straight line across the playground.
Get a friend to follow you and record your path with chalk.
Begin at the same point, this time blindfolded.
Try again, with another friend beside you to make sure you don't bump into anything!
Record the path again and compare the two lines.
Are girls or boys better at this?

Feeling with Your Toes

Both partners take off their socks and shoes.
Blindfold one partner at a time.
Touch different objects first to the feet and then to the hands of your partner.
Can s/he guess what each is just by feeling it with the toes?
Can s/he guess just by feeling it with the hands?
Change the blindfold and repeat the experiment.
Make a chart to show the results for both of you.

Try to paint a picture of a tree—with the brush held between your toes!!!
Can you do it?

Guess Sounds and Direction

Choose a large and quiet area.
Blindfold your partner, who must sit in a chair, head pointed forward.
Drop different objects on a table.
(e.g., a knife, a coin, an eraser, etc.)
Can your partner guess what they are?
Record the guesses.
Change the blindfold and repeat the experiment.

Try it with several different coins.
Do either of you get better with practice?

Ring a bell at chest level, eye level, to the side, etc.
Your partner should try to point to the bell.
Record the correct guesses and the direction.

Guess That Sound

Record various sounds and play the tape to the group.
Who can guess the sounds?

Record different student's voices.
Who can guess whose they are?

Hair and Eyes – Who Has What?

Make a chart to record everyone's hair type and color.

Name	Color of Skin *(dark, light)*	Color of Eyes *(green, grey, blue hazel, brown)*	Color of Hair *(black, blond, brown, auburn, gray, white)*	Type of Hair *(straight, curly, wavy thick, fine)*

Do many children look alike?
Do people of different skin color tend to have similar hair and eye color?

Measure Your Feet and Your Hands

Trace around the feet of your partner on newsprint (with your shoes off!).
Cut out the prints.
Measure the length from the heel to the big toe.

Trace around your hands.
(Spread your fingers out before tracing.)
Measure from the tip of the thumb to the tip of the little finger.

Glue all the class's feet and hands to a large sheet of paper from the smallest to the largest, with the names written underneath.

Jumping Up

Get some chalk dust on your fingers.

Stand flat beside the blackboard and stretch your arm to mark the board.
Jump as high as you can and make a second mark.
What is the distance between the two marks?

How Much Can You Pick Up?

Fill a large bowl with seeds.

Pick up as many as you can with your right hand. Count them.
Try again with your left hand. Count them again.
Draw a chart to compare your count with your friends.

LOOKING AT AN APPLE

ACTIVITIES FOR THE STUDY OF MATTER (K-3)

UNIT OBJECTIVES

Skills

Observation

* Study the taste, feel, sound, and smell of an apple.

(While all the senses should be used in observation, their contribution varies with the nature of the object being investigated.)

Inferring

* Make inferences to fit observations as to differences among varieties of a fruit.

* Make inferences to fit observations of the behavior of a non-symmetrical shape when examined by a variety of means.

* Become committed to an outcome before actually doing an activity.

Communication

* Describe the different varieties of apples.

* Draw pictures, describe, share, compare, classify.

Attitudes

Curiosity

* Use several varieties of apples and alternate foods in your explorations.

Concepts

Change

* Look at an apple before and after cutting.

Interaction

* When we handle a substance, we change it.

* Consider the apple as part of a food chain.

Unit Activities

Examine an Apple Carefully

What color is the apple?
Is it the same color all over?
Are all apples the same color?
What different colors are they?
Does the apple look smooth, lumpy, rough?
What happens to the color when you polish an apple?

Look at an Apple

EXAMINE THE SHAPE OF AN APPLE.

Place an apple on its side and roll it.
Roll it on its base.

What shape is the apple?
What things are the same shape?
In what position did it roll most smoothly?

Stand an apple on its stalk end.
Use pieces of wool to find the perimeter along the horizontal and the vertical axes.
Which is longest?
Measure a tennis ball the same way and compare the results.

Listen to an Apple

SHAKE, RUB AND ROLL AN APPLE.

Can the seeds be heard to rattle?
What other surface gives a similar sound to that made when an apple is rubbed?
What sound is made when the apple is rolled?

Smell an Apple

SMELL AN APPLE BEFORE AND AFTER IT'S BEEN CUT.

Is there a smell on both occasions?
At which time is the smell stronger?
What words can you use to describe the smell?

COMPARE THE SMELL OF APPLES OF DIFFERENT VARIETIES AND SIZES.

Does the size of the apple affect the strength of the smell?
Do red and green apples smell the same or different?
If an apple gives off odor will its mass change?
If an apple that has been cut is left for several days, is the odor different? Is it stronger?

Distinguish apples from other objects by odor.

Draw a picture of the part of the apple that has the strongest odor.

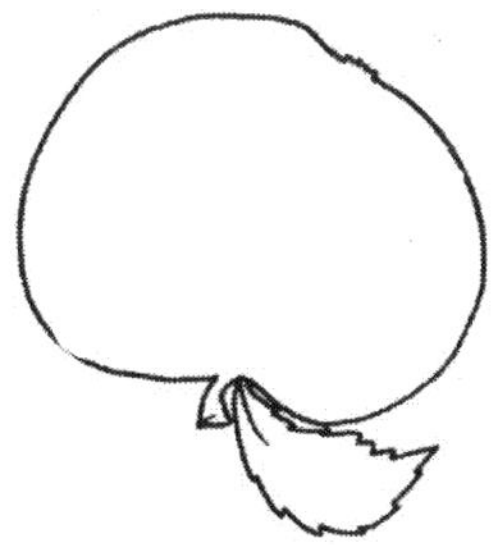

Feel an Apple

ROLL AN APPLE BETWEEN THE PALMS OF YOUR HANDS, AND THEN TRACE THE OUTLINE WITH YOUR FINGERS.

Does the apple feel most like a sweater, a bottle, a colored rod?
What things does it feel like?
Is it smooth, cool, round, hard, dry, small, soft, springy?
Does it feel cool, warm, hot?
Is it more or less slippery than a desk top, an orange, plasticene?
Do all apples have a similar texture?
Draw a group of objects with similar texture to an apple.

Taste an Apple

EXAMINE THE CUT APPLE CLOSELY.

Do they all taste alike?
What kind of apple is the crispest, the sweetest, the juiciest, the sourest?
Are big apples sweeter than small apples?
Which fruits taste sweeter than apples?
Is any other fruit as crisp as an apple?
Does a little sugar, or salt, change the apple's taste?

COMPARE APPLES OF DIFFERENT VARIETIES. LISTEN TO THE SOUNDS.

Is the sound different when you or someone else chews?
Do different varieties of apples give different sounds?
Do apples that have been kept in the refrigerator sound different?

What sounds are made when oranges, bananas or cookies are eaten?
Which is most like an apple?

TASTE PIECES FROM SEVERAL APPLES OF DIFFERENT VARIETIES.

What color is the skin and the flesh of the apple?
Does cutting change the skin?
How thick is the skin?
Are the cut surfaces damp?
Where are the seeds found?
How are the seeds held in the apple?
Examine cut apples closely.

LIGHT

ACTIVITIES FOR THE STUDY OF ENERGY (K-3)

UNIT OBJECTIVES

Skills

Classifying

Categorize sources of light.

* Group objects according to the degree of reflectivity and differences in absorption.

Space/Time Relationships

* Examine the progress of day and night at different times of the year.

* Describe the sequence of an experiment.

Attitudes

Persistence

* Complete all stages of a multi-stage experiment.

* Follow written instructions without constant need for assistance.

Inventiveness

* Devise way of examining the spectrum.

* Examine and test several different shiny and dull objects.

Concepts

Interdependence

* Light can be changed so that it affects the color of objects.

* Light in a dark place may be increased by the use of an object that either gives light or is a good reflector of light.

Variety/Commonality

* Some materials absorb light, some reflect light, some stop light, and some let light through.

* There is a relationship between light and the colors of the spectrum.

UNIT ACTIVITIES

Think About Light

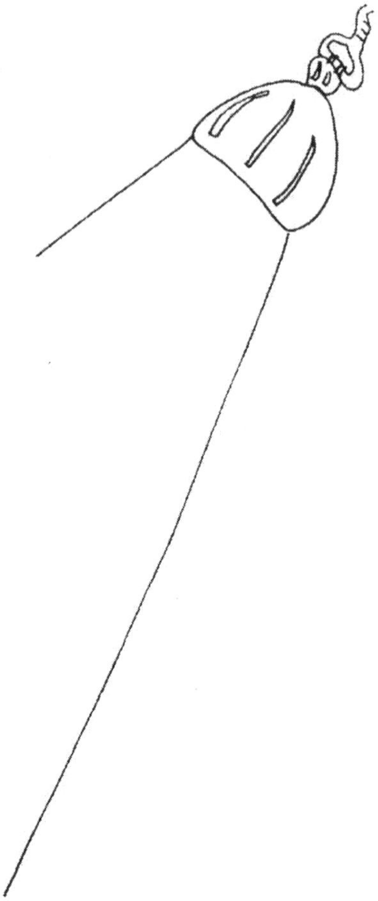

List as many sources of light as you can, including natural sources and man-made sources.
What is our planet's main source of light?
Do we get light from the sun at night as well as during the day?
What things have people invented and made so that we can have light when and where we need it?
When might some of the man-made sources of light be used?
What things can we see only when light from something else is shining on them?
What things make light,
so that we can see them even if it is very dark?

Darken a Room

You will need a lighted candle or other source of light, a white article such as a sheet of paper, and a
shiny object such as a spoon or mirror.
What objects are easy to see?
(The candle? Lightly colored, or dull-colored objects? Shiny articles?)
Blow the candle out. What objects are easy to see now?
Are some pieces of furniture easier to see than others?

How to Be Seen at Night

Two people walk around in a darkened room—one in light clothes and one in dark clothes.
Which person can be seen when there is no light?
Which can be seen when a flashlight is shone around the room?
How close do you need to be to see someone wearing dark clothes on the street on a dark night?
If they have light clothes on?
Why might the driver not see you even if you can see the car lights?

Look into a Box

Cut a slit in a closed cardboard box and look into it.
What can be seen? The color of the inside? The joins or corners? Nothing at all?
Put a lighted flashlight into the box. What can be seen now?
Remove the light and place a dull and a white object in the box.
Which object is easier to see?
Find different shiny and dull objects to test.
How does the degree of shininess or dullness combined with a small amount of light affect the visibility of things?

Look for Materials that Absorb or Transmit Light

Hold a lighted flashlight so that the light image can be clearly seen on the blackboard when the room is slightly darkened.
Watch the image closely while a variety of materials are placed over the flashlight glass.
Which materials let the light through?
Can you see through these things easily?
Are there materials that stop the light?
Can you see through these materials?
Are there some materials that let through some light, but not enough to show on the blackboard?

Examine the Spectrum

Look at rays of sunshine coming between trees or through a window.
What color are they?
Can any other colors be seen?
Blow bubbles where they can float through light rays and watch them carefully.
What do you notice about their color?
Where else are colors like this seen?
What do you notice about the color of the bubbles when they are blown in a shady place?
Add detergent to a bowl of water placed in the rays of sunshine and beat it to make a mass of bubbles.
Do bubbles cause colors to appear in the water?
Are these colors like rainbows?
Move the bowl into a shady or a darker place.
What happens to the colors?
Find other ways to make rainbow colors.

Making a Spectrum

Set up a large jar filled with water and place it on a table in the path of direct sunlight.
Place a large sheet of white paper on the floor.
Move it around until the spectrum appears.
Describe the colors seen.
Describe the shapes that the colors make on the paper.
(e.g., circular, splotchy, in lines, etc.)
Shift the paper closer to the jar. What happens?
Tilt the jar. Does the position of the spectrum change on the paper?
Block the sunlight. Do the colors still appear?

Filter Light

Examine scenes or objects through colored, translucent material such as cellophane or plastic.
How is the color of the objects affected?

Which colors are not affected?
Are they colors that are like the cellophane or are they different?

Put a colored, translucent material over a flashlight and shine the light onto objects.
How is the color of the objects affected?

In what other ways is the color of light changed?

Stare at a green light or a bright green object.
Close your eyes. What color do you see?

NOTES FOR THE TEACHER

Natural light sources: stars, electricity, moon, fire, etc.
Electrical light sources: street lights, car lights, home lighting, etc.
Chemical light sources: luminous tape, figures on a watch, etc.

Changing the color of light: colored glass used in neon and party lights, tail indicators on cars, traffic lights, etc.

The activity that requires the students to stare at a green light is a good, practical introduction to complementary colors in art.

THE LIVING EARTH

ACTIVITIES FOR THE STUDY OF THE EARTH (K-3)

UNIT OBJECTIVES

Skills

Predicting

* Predict animal behavior based on observations.

* Predict the weather based on observations.

Controlling Variables

* Examine wind in specific locations.

* Examine animals in controlled environments.

Interpreting Data

* Examine weather reports in the paper, online, or on T.V..

Attitudes

Caring

* Show a respect for living things and their environment.

Critical Thinking

* Examine how different parts of the earth interrelate.

Concepts

Conservation

* Recognize ways in which we can protect our environment.

Continuity

* We can observe many cycles in the weather.

UNIT ACTIVITIES

Think About the Physical Earth

Examine Sand

Examine sand closely through a strong magnifying glass.
Look for differences in color, shape and size.

Colorless glassy crystals are quartz.
Red crystals are usually garnet.
Flaky black crystals could be mica or biotite.
Rectangular black crystals are most likely hornblende.
If any black crystals are attracted to a magnet, they are magnetite.
Green crystals may be olivine.
Purple crystals may be amethyst.

Observe Sediments

In a large glass jar, add small stones, sand, and fine earth, until the jar is 1/3 to 1/2 full.
Add water until the jar is almost full.
Screw on the lid and shake vigorously.
Put the jar down and allow the contents to settle.

Heavy, course particles will settle out first.
The lighter, finer particles will form a layer on the top.

Form Crystals

Dissolve as much sugar as possible in a glass half-filled with hot water, stirring vigorously.
Pour the solution into a small, deep saucer.
Put a string in the solution, with the end hanging over the edge.
Allow the solution to evaporate for a day or two.
Pour off any solution that remains, allow the crystals to dry,
and examine the crystals through a magnifying glass.

Repeat the experiment, using borax, alum, and salt.
Compare the crystals in each case.

Examine Folding

Fold three differently colored towels lengthwise.
Place them on top of one another to represent layers of rock.
Place one hand near each end of the pile of towels, and push sideways, toward the middle.
Examine the wavelike mountains and valleys.

Think About Environments

Find and Examine Animal Homes

Look under grasses, in bushes, under rocks, in river banks.
Look in the sand and in shallow water.
Describe the homes.
What kinds of plants are the homes near?
Consider why the homes are where they are.

Look for Patterns of Animal Behavior

Do they sleep at night or in the day?
How far from home do they travel?
How do they protect, or disguise, their homes?
Do they live in groups or alone?
What do they eat to stay alive?
Do they live near the plants or animals that they like to eat?

Examine Plant Locations

Look for differences in plants that are near, or far from, water.
Look for plants that grow alone, and those that grow near trees.

Create an Environment

In a box or aquarium tank make a model
of one of the environment that you have studied.

Collect some water from a lake, river, stream, or the ocean.
Examine some of the water under a microscope.

Collect a boxful of earth from a vacant lot or wooded area.
Examine the dirt for plants and animals.
Consider how they relate to each other.

FORMS OF LIFE			ENVIRONMENTS
grasses trees bushes flowers	insects worms birds animals	fishes shellfish mollusks amphibians	bodies of water marshes rivers, streams, and creeks forested or wooded areas hills and mountains

Keep insects only long enough to study them a little bit.
Always release animal life back into the environment it was taken from
to avoid cross-contamination of environments.

Think About the Weather

Examine Rain

What kinds of clouds bring rain?
What different kinds of "rain" are there?
Why do insects bite more before a rain?
Why do some blossoms close up before a rain?
Why do leaves on some trees show their undersides?
Why are children more excited in a thunderstorm
but quieter during ordinary rain?
Why do older people complain of aching joints during a storm?

Create Rain

Boil some water in a saucepan.
Hold a metal tray of ice over the steam.
What happens as the steam cools a bit?
What happens as it cools a lot?

Examine Wind

Find out about the wind patterns around your school.
Where is the best place near your house to fly a kite? Why?

Use a variety of objects to move the air in the room around.
Heat some air and look through it.
Hang a pinwheel above the heated air.

Examine Thunder and Lightning

Use two long balloons to create sparks.
Rub them on your clothing or a rug.
Hold them very close together in a dark room.
Imagine a strong wind "rubbing" the water drops in the clouds.

Use a paper bag to create thunder.
Blow the bag up and hold the end tightly.
"Pop" the bag.
Describe the effect.

NOTES FOR THE TEACHER:

- When water is heated, it evaporates. When cooled, it condenses. This condensation forms a cloud of rain droplets. When the drops are cooled and get heavy, they fall as rain.
- Wind is the movement of air set up by the unequal heating of the earth's surface by the sun.
- Lightning is a huge spark of electricity.
- The air in the bag will be compressed. It expands rapidly as it escapes, creating a sound wave that reaches your ears as a pop. Thunder is the sound of air exploding as lightening heats it up, creating a sound wave of vibrating air.

FLIGHT

ACTIVITIES FOR THE STUDY OF SPACE (K-3)

UNIT OBJECTIVES

Skills

Predicting

* Predict the outcome of various experiments prior to doing the activities.
* Use the results of activities with seeds to predict the flight of a hot-air balloon.

Designing Experiments and Controlling Variables

* Ensure that all experiments are done in a scientific manner, with accurate measurements, care that the variables are examined one at a time, and carefully recorded results from which conclusions can be drawn.

* Plan and design an experiment to study some aspect of space exploration.

Attitudes

Curiosity

* Examine many different ways of learning about flight.

Rationality

* Distinguish between folklore and scientific discovery in the study of flight.

Concepts

Cause and Effect

* Technological growth causes, and is caused by, scientific development.

Integration

* The study of flight and space crosses boundaries of all the content areas: animals, plants, matter, energy, earth, space, health, and technology.

Unit Activities
Principles of Flight: 1) Lift

Examine how air flows over and around an airplane wing.

Hold a piece of paper just below your lower lip.
Blow across the top of the paper and see how the paper rises.

Suspend a length of paper loosely between two piles of books.
Blow across the top of the paper.
Does the paper rise?

Suspend two sheets of paper about one inch apart between two stacks of books.
Blow between the sheets of paper.
What happens to the paper?

Use lightweight string or thread to suspend two table-tennis balls about one inch apart.
Using a straw, blow between the balls.
What happens to the balls?

Suspend a table tennis ball in the stream of air formed by a hair dryer or a vacuum cleaner with its hose attached to the blower end.
Draw a picture of the results.

Place a table tennis ball in a glass funnel.
Attempt to blow the ball out of the funnel.
Try it with the funnel facing down after you start blowing.
What happens?

Suspend two spoons, bowls back to back and slightly separated.
Direct a flow of water from a faucet between them.
What happens to the spoons?

Tape or tack a card to a ruler so that it curves like an airplane wing.
Balance the ruler on a 6-sided pencil with the card end slightly lower.
Using a straw, blow across the curved surface of the card.
What happens to the card and ruler?

Pull a strip of paper through the air.
What happens to the paper?

Push a sheet of paper forward through the air.
What happens to the paper?

The movement of air, or water, across a surface reduces the pressure, so pressure is increased on the opposite surface.

Principles of Flight: 2) Gravity

Observe how things fall in different circumstances.

Jump into the air.
Why did you come down again?

Drop a pencil, a rock, and a ball.
Why did they come down again?

What would happen without gravity?

Jump rope.
Why can the rope be "turned"?

Place marbles in a container suspended by a rubber band or a spring.
Add extra marbles.
How does the increase in the length of the rubber band or spring indicate the increased pull of gravity?

Principles of Flight: 3) Thrust

Examine ways in which we can get things moving.

Place a fan on a board which is lying on round pencils or dowels.
Observe the movement of the board as the fan is turned on.

Place the fan in a small wagon.
Observe what happens when the fan is turned on.

Fly a model airplane with a propeller powered by a rubber band.

Tape an inflated balloon to a straw.
Thread a long, lightweight wire through the straw and hold or fasten it at each end so that it will be taut.
Release the air from the balloon and observe its movement along the wire.

Fasten an inflated balloon to a small, lightweight toy so that when air is released, the toy will move forward.

Tape an inflated balloon to a model airplane.
Release the air and observe its "flight".

Principles of Flight: 4) Drag

Observe the forces that can keep a plane from flying.

Run with a large square of light cardboard held flat against the wind.
Repeat the run with the edge against the wind.
Describe the difference in the amount of resistance.

Open an umbrella.
Hold it high above your head and pull it down quickly.
Describe how the air resists the movement of the umbrella.

Drag a clothespin through a bucket of water.
Attach a ketchup bottle cap with a string and drag it again.
Describe the difference in the two experiments.

Drop two identical sheets of paper together: one flat, one crumpled.
Observe the effect of air resistance on the rate of fall.

Drop a sheet of aluminum foil and a marble at the same time.
Describe what the difference is.

Crumple the foil into a ball and again drop the foil and marble at the same time.
Describe what the difference is this time.

Make a parachute of a large handkerchief.
Tie an object to it.
Drop an identical object at the same time as the object fastened to the parachute is dropped.
Describe how the air resistance slows down the fall of the object attached to the parachute.

High-Altitude Flight

Try eating and drinking while standing on your head.
Discuss the effect of weightlessness on the ability to eat and drink.

The feeling of weightlessness may be experienced to a small degree on a playground swing.
Notice the light feeling at the top of the swing's arc just before it begins its swing back toward Earth.
Notice the feeling of heaviness as the swing passes close to Earth.

Discuss the feelings experienced in riding down in a fast elevator,
riding in a car or on a bike over a large bump in the street,
jumping off a diving board, or riding on a roller coaster.
You can compare these experiences to weightlessness.

PENDULUMS

ACTIVITIES FOR THE STUDY OF TECHNOLOGY (K-3)

UNIT OBJECTIVES

Skills

Measurement

* Accurately measure the length of strings used in the experiments.

* Measure the differences caused by changing components.

Controlling Variables

* Ensure that the experiments are carefully set up so that results can be measured as accurately as possible.

Designing Experiments

* Try to develop some original experiments based upon the results of classroom activities.

Attitudes

Curiosity

* Be prepared to learn why objects behave in a particular fashion.

* Examine the effect of changing various parts of the pendulums.

Inventiveness

* Consider many different ways to develop an understanding of the motion of pendulums.

Concepts

Interaction

* Gravity affects the motion of all things.

Cause and Effect

* Changes in the motion of pendulums can be related to changes in the experimental set-up.

UNIT ACTIVITIES

Making Pendulums

MAKE A PENDULUM BY HANGING A WEIGHT (A BOB) FROM A PENDULUM SUPPORT.

Make a note of the length of the string.
Pull the bob back and hold it.
How far do you think it will swing to the other side?
Hold your hand where you think the bob will reach.
Now, let the bob go.
Were you right?

Count the number of swings the pendulum makes before stopping.
Which swings out farther, a heavy bob or a light bob?
Which swings for a longer amount of time, a heavy bob or a light bob?

Hand Pendulums

TRY SOME ACTIVITIES WITH A HAND-HELD PENDULUM.

When the pendulum is not swinging, how can you move your hand to start it swinging?
In how many directions can you move your hand and keep the pendulum swinging?
Can you make the pendulum swing very fast?
Can you move your hand without moving the pendulum?
Can you stop the pendulum from swinging just by moving your hand?

Pendulums Are Swingers!

DRAW ALL THE WAYS YOU THINK A BOB CAN BE SWUNG.

How else can you swing a bob besides straight, back and forth?
With a pencil close to the bob, follow it as it swings, tracing a line on a piece of paper showing how the bob swung.
Can you make the bob go in a circle?
...an egg shape?
...a triangle?
What other shapes can the pendulums swing make?

Get It Together!

Can you get two bobs of unequal weight to swing together for 10 swings?

Adjust the length of the string for two pendulums of equal weight so that one swings 20 times in the same amount of time the other swings 10 times.
Exchange one of the bobs with very heavy bob.
Will the ratio still be 20 swings for 10?

Can you get two pendulums to swing together
...for 10 swings?
...for 20 swings?
Try to get three pendulums to swing together
...for 10 swings.
...for 20 swings.
Try to get odd shapes to swing together.

Criss-Cross Pendulums

Can you swing two adjacent pendulums—one back-and-forth, the other side to side—so that they do not hit for 10 swings?

Can you do it for 20 swings?
Try this with two pendulums hung side by side and swinging in opposite circles—one clockwise, one counter-clockwise.

Pendulums in Orbit

1) SET UP 20 TEES IN A SQUARE.

T T T T T
T T T T T
T T T T T
T T T T T

Swing the pendulum in orbit around the outside of the group of tees.
After three orbits can you predict which tee will be last to be knocked over by the bob?
(You may change your prediction only once.)

2) SPREAD OUT 20 TEES LOOSELY BENEATH A PENDULUM.

Set the pendulum in orbit around the outside of the group of tees.
(The orbit is to prevent the bob from hitting the tees.)
Watch the circling bob.
If it looks as if the bob will hit a tee, pick it up and place it somewhere else in the group, where you think it won't get hit.
When the pendulum stops, look closely at the position of the tees.

Pendulum Bowling

Set up ten golf tees in a triangular arrangement.

T T T T
T T T
T T
T

Use a small bob.
How many "pins" can you knock over with three swings?
Record your score for each three-swing attempt.

Change to a heavy bob.
Change to a light bob.

Play the game with a hand-held pendulum.

Salt Pendulums

MAKE A SALT PENDULUM.

Cut a small hole (just large enough for salt or sand to flow through) in the bottom of a paper cup.
Hang the paper cup from the pendulum support, in any way you like.
Fill the cup with salt (or sand) and swing the pendulum.
Look at the pattern made with the salt.
Draw the patterns that you made with the salt pendulum.

MAKE A COMPOUND SALT PENDULUM.

Hang the cup from a Y string.
The knot at the junction of the string Y should be a slip knot so you can move it up or down.

Swing the pendulum with the junction in different positions.
Note the different patterns the salt will make.
If you wish to make a permanent pattern, put glue on the paper first.

With a partner, try some pendulum activities on the swings.

Try the "salt" experiments again as you swing.

SECTION 8B - ACTIVITIES FOR GRADES 4-7

ANIMALS IN A MICRO-HABITAT

PROPAGATING PLANTS

PHYSIOLOGY

LOOKING AT BUBBLES

VIBRATIONS

THE LIVING EARTH

EXPLORING THE MOON

FORCE, FRICTION, AND MACHINES

ANIMALS IN A MICRO-HABITAT

ACTIVITIES FOR THE STUDY OF ANIMALS (4-7)

UNIT OBJECTIVES

Skills

Observation

* Study animals in a variety of habitats.
* Study various animals of a specific habitat.

(While all the senses should be used in observation, their contribution varies with the nature of the object being investigated.)

Inferring

* Make inferences to fit observations as to how animals obtain water, air, shelter, and protection.
* Develop divergent thinking skills by suggesting several inferences to explain an observation or event.

Communication

* Describe the variety of animals and habitats found.
* Draw pictures, describe, compare, classify.

Attitudes

Curiosity

* Examine a variety of animals and environments.

Caring

* Handle small animals with care.
* Devise adequate micro-habitats for the classroom.

Concepts

Change

* Animals may change their environment to suit their needs.
* Animals will adapt when moved to different habitats.

Interaction

* Animals are dependent upon their environment for survival.
* Some animals construct shelters, others may use natural cover.

REMEMBER, DO NOT DO ANYTHING THAT WILL HARM YOUR ANIMAL!

Unit Activities

Planning the Collection of Animals

Look for animals in an area which has not been sprayed, mown, or kept pruned.

The best areas to consider are overgrown lots, areas where low bushes are growing, and areas where there is rotting timber or stones among the plants.

Select and mark out a small area (micro-habitat) in the local surroundings.

Will the animals collected in different ways be the same or different?

Does the animal need water?
How does it obtain water in its natural environment?

Planning Your Temporary Environment

Consider the types of homes that need to be prepared for any small animals which may be found.

A shallow dish of water with a cloth wick to conduct water into the soil and atmosphere of a container will improve the temporary environment for most insects and small animals, such as ladybugs, caterpillars, snails, slugs, grasshoppers, lizards, frogs, garden spiders, or crickets.

Consider how to keep the animal comfortable.
Does it seem to want to keep out of the light?
Was it hidden under the leaves when you found it?

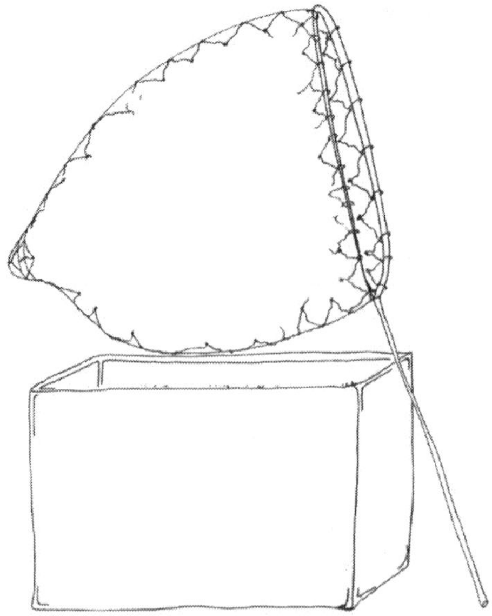

Collecting Animals from Grassy Areas

Use a stocking net to sweep the area quickly several times, rather than moving the net slowly.
Handle the net carefully after each sweep even though it may appear to be empty.
Have containers ready for any creatures found.

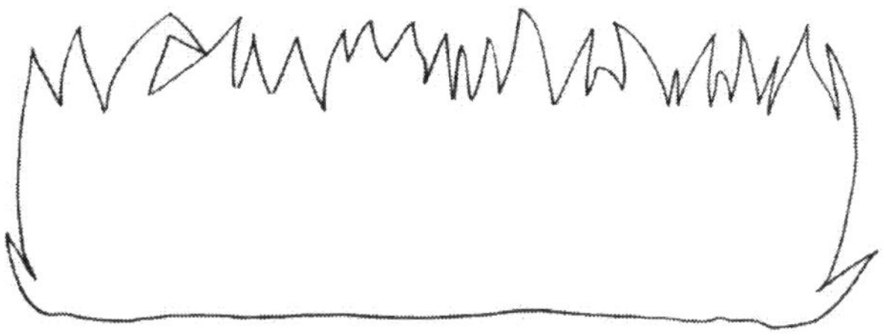

Collecting Animals from Around Roots and Under Logs

Consider whether there will be any legless animals in this environment.
Predict what sort of food these animals might eat.
Use a trowel and strainer to dig and sift soil
from around the roots of plants and under rocks or logs.

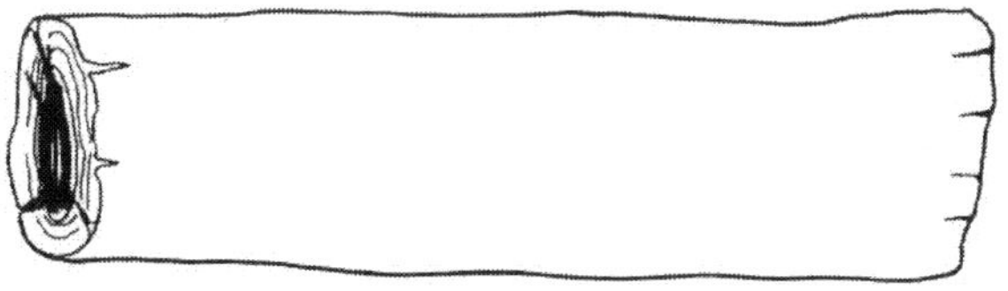

Collecting Animals from Bushes

Place a large sheet of cloth or plastic under bushes, and shake the branches
to unsettle animals sheltering in the leaves.
Observe and make inferences about how these animals are like/different
from those found in grass or under the soil.
Observe the number of winged animals found living in bushes.

Collecting Animals from the Ground

Sink a tin into the ground so that its open top is level with the ground.
Bait the can with different foods.
Prop a piece of bark on a rock to keep the tin dry and accessible.
Discuss the animals caught in the "trap".
Consider the types of animals that fall into the trap
during the day, during the night, or at either time.
Consider whether certain kinds of "bait" attract certain types of animals.
(sugar, meat, pieces of fruit, etc.)

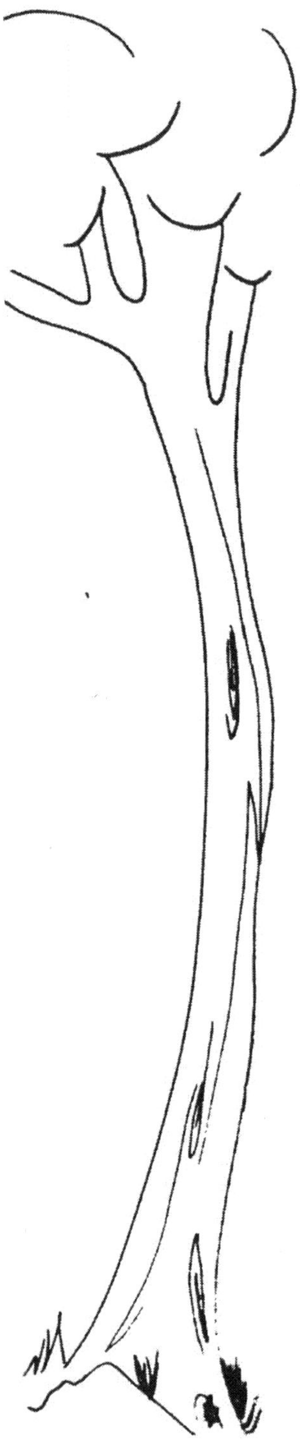

Collecting Animals from Trees

Find different animals that live in the bark of trees.
Examine the leaves of trees to look for small animals.
Examine the litter at the base of the trees.
Look for "homes" in trees, such as cocoons, nests, webs, and hives.

Identify and Classify the Animals Collected

Study the animal closely using a magnifying glass, hand lens, insect magnifier, or stereo-microscope to observe details in structure.

LOOK FOR CHARACTERISTICS.

Has the animal legs, wings, feelers (antennae), eyes...?
How many legs, wings, eyes can be distinguished?
Can separate body parts be distinguished? How many?

Look for mouth parts.
Can pincers, tongue, jaws be seen?

Are the wings clear or opaque?
When there are two pairs of wings, are they alike?

Are all the legs similar?

GROUP THE LIFE YOU FOUND.

spiders
insects without wings
insects with wings
young insects
animals with more than eight legs
animals with no legs
pests vs helpful creatures

PROPAGATING PLANTS

ACTIVITIES FOR THE STUDY OF PLANTS (4-7)

UNIT OBJECTIVES

Skills

Measuring

* Chart the growth of different plants.

* Examine the ratios of size of different parts of a plant.

Classifying

* Examine differences in plant growth according to plant type and plant part.

* Recognize the different parts of several plants.

Space / Time Relationships

* Explain the sequence of development of growth in a plant.

Attitudes

Persistence

* Carry through the propagation of a plant from preparation through planting and growth.

Inventiveness

* Consider a variety of plant comparisons.
* Prepare a garden.

Concepts

Interdependence

* Plants respond to stimuli from the physical and living environment.

Variety/Commonality

* Plants grow and change at different rates.

* When plants are propagated, the new plant has the same characteristics as the old plant.

* Plants of the same species may vary in size and appearance under different environmental conditions.

UNIT ACTIVITIES

Study a Plant

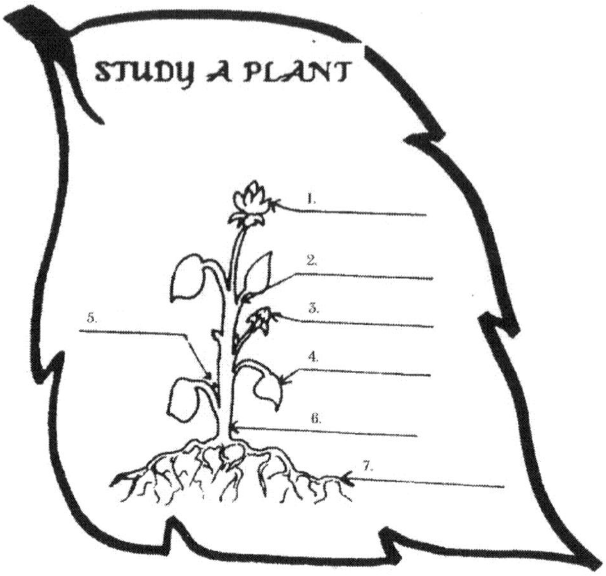

Label the parts of the plant diagram.

Examine Plants

Use a hand lens to study an actual plant and to sketch the parts observed.

Collect, cut, and compare:
- rhizomes and runners
- corms, bulbs, and tubers

Plant Different Plant Parts

Collect parts of plants to investigate which parts grow new plants.
Record your results in a chart.
Sketch the plants that grew successfully, indicating the part used.
Discuss reasons for observed differences.

Propagate Sweet Potatoes

Examine the skin of a sweet potato carefully. Describe what you see.

Put toothpicks into the sides of the sweet potato and, balancing the toothpicks on the rim of a glass, submerge its base in a glass of water.
Examine where and when roots and shoots first appear.
Describe the shoots and leaves at early and later stages.
Repeat the experiment with an onion and compare the results.

Prepare Cuttings

Cut the stem of a plant below a node.
Remove the lower leaves and plant it in a flower pot.

Prepare cuttings from a variety of different plants
(hibiscus, geranium, rubber tree, willow, etc.).
to investigate types of plants which grow well
from cuttings under normal conditions.

Select three cuttings of approximately the same size
from the same plant to investigate
the conditions under which they grow best.
Place one in water, one in moist soil,
and one in moist soil covered with a plastic bag.

Try to control all the other variables
(e.g., the amount of sunlight, exposure to wind, etc.).

Determine what conditions best suit different types of cuttings.

Make a Garden

Prepare a garden in a saucer, bottle, rockery, aquarium, etc.
Record what medium you are growing the plant in.
(E.g., soil, water, cotton wool, etc.)
Record the names of the plants and the parts used.
Record the growth rate of different plants.
Describe each plant's water needs.

NOTES FOR THE TEACHER

The term PROPAGATION includes all the methods by which plants can reproduce themselves.

VEGETATIVE REPRODUCTION occurs when we use:

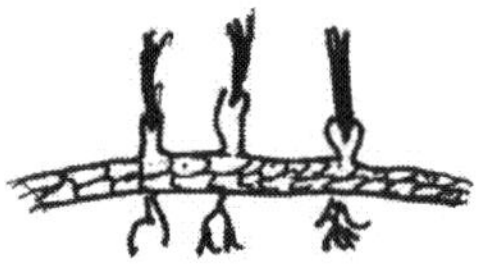

Rhizomes: underground stems (horizontal) which bears buds in the axis of reduced scale-like leaves (e.g., bamboo, couch, etc.)

Stolons (runners): horizontal above ground stems which root at the nodes, forming new plants that eventually are freed from connection with the parent when the runner decays (e.g., strawberry, buffalo grass, violet, etc.)

Corms:: swollen stem bases containing a store of food material and bearing buds in the axis of scale- like remains of leaves of the previous season's growth (e.g., gladioli, anemones, etc.)

Tubers: swollen, food storing roots with visible "eyes" (e.g., potatoes, dahalia, etc.)

Leaves: often fleshy leaves (e.g., begonias, african violets, succulents, etc.)

Bulbs: modified stems enclosed by fleshy, scale-like leaves (e.g., onion, daffodil, etc.)

PREPARING CUTTINGS – Cutting of certain plants normally root slowly or produce a very small number of roots. It is possible to hasten the root process by:

- treating the cutting with rooting powder (auxins)
- leaving a few leaves on stem cuttings, not only because they contribute carbohydrates to the roots, but also because they may contribute hormones.

Willows grow best in water, geraniums in soil.

Study a Plant (Diagram)

1. flower; 2. petiole; 3. terminal bud; 4. leaf; 5. auxiliary bud; 6. stem; 7. roots

PHYSIOLOGY

ACTIVITIES FOR THE STUDY OF HEALTH (4-7)

UNIT OBJECTIVES

Skills

* As an integrated unit, the following does not lend itself to the definition of specific process skills development. Rather, the thinking skills are developed through the use of creativity and decision-making activities.

Attitudes

Caring

* Examine emotional and social aspects of the human condition.

Critical Thinking

* Consider controversial issues in the study of the human body.

Concepts

Continuity

* The many parts of the human body are interrelated to insure the continuity of life within an individual.

* * *

UNIT ACTIVITIES

Artistic Creativity

You are a travel agent taking microbes on tours through the human body.
Design a catalog of places that your clients could visit.
(e.g., heart, liver, lungs, ear, eye, tongue, skin, etc.)

Outline travel arrangements and points of interest.
Point out the potential hazards, and the precautions that must be taken.
Make sure that your catalog is colorful, interesting, and accurate.

➥

Choose one of the body's systems, and design a Coat of Arms.
Consider factors such as function, importance, "social standing".

➥

Make up an Organ Catalog for a Clone Factory.
Be sure to give the uses for each organ.

➥

Make up an advertising campaign for your favourite food.
Make a poster, some badges, and a speech!
(ALTERNATIVES: Organ of the Month, Muscle of the Month, etc.)

➥

Make a mobile of drawings of the various organs found in the human body.

Written Creativity

Describe HEALTH.
Write all the words you can think of that describe a healthy person.
Now write a list of descriptive words for an unhealthy person.

➥

You are a white blood cell on the trail of a germ.
Write your report to control center, explaining how you will find out where the germ is, how you will track it down, and what you will do to it.

➥

You are going on a visit to another planet.
Prepare a medical folder that you can take with you that will explain to an alien doctor how it can care for you if you get sick.

➥

Write up a union bargaining plan for the autonomic nervous system as it is trying to develop a new contract with your body.
Keep in mind hours of work, working conditions, and so on.
Does your body do things to make it easier or harder for the nerves to do their work?
(e.g., types of stress, general physical condition)
What could you offer your nervous system?

➥

Create a newspaper for the human body.

News—written by the brain
Weather—written by the skin
Sports—written by the muscles
Market News—written by the stomach

➥

Write a five day diary in the life of: a heart, a foot, an eye, etc.

➥

Make up a book of jokes or riddles about: human systems, nutrition, fitness, emotions.
Be sure that the jokes tell us something and that they are not tasteless or rude.

➥

Make up a poetry book to describe one of the following: your senses, your feelings, nutrition, fitness.

➥

You are building a human body.
Write job descriptions for five organs that you plan to hire to work in the body.
(ALTERNATIVES: systems, glands, bones, muscles, etc.)

➥

Examine a special event from the point of view of your body.
For example, how does your body prepare for, and recover from, a birthday party?

EMOTIONALLY: Are you excited or nervous?
How does your body react to all this?
PHYSICALLY: Is your stomach ready for the extra food?
Did you stay up late and get overtired?

➥

Choose a system of the human body and write a "Maintainance and Repair" booklet.

➥

Write a "Letter to the Editor" from the reflex system explaining the stress that it is under.
(e.g., it must always be ready to act, must be instantly operational, etc.)

Decision Making

Choose one of the human systems.
What "laws" must it obey?

➥

Give 10 reasons for eating well.
Give 10 effects of eating poorly.

➥

How would you like living in a society where you would only have to take a pill or two to get all your food requirements?

➥

What do you think are the best ways to cope with pimples or acne?
How can you help prevent it?
How can you help cure it?

➥

What might be difficult about being one of the following:
food, an organ, a system, a gland, etc.

➥

Explain why we are so embarassed when we discuss the reproductive or excretory systems.

➥

Argue the question: *The human body has too many organs.*

➥

Choose to be a part of the body.
(e.g., the heart, toe, brain, pancreas, nose, etc.)
Decide what things are important to you in your life.

➥

Write a short essay explaining why it is important to have a warm-up before doing something very strenuous, such as skiing, playing hockey, etc.

➥

Should girls in elementary school wear make-up?
Have the skin and the emotions argue the question.

➥

Discuss the value and the problems of transplants.
Examine social, physical, emotional, and ethical considerations.

➥

Write a short essay outlining how you think you'd cope with a crippling injury or disease.
OR
Write a short essay outlining how you think you'd cope with a serious operation.
(There might be quite a bit of pain over a long period of time.)
Consider social, emotional, and physical aspects.

➥

Your job is to regulate all traffic in the human mouth.
What traffic signals are there?
What facilities can we find?
How do you keep the traffic flowing smoothly?
How do you detect and stop contraband?
What laws would you like to see in force?

Research

Choose one of the systems of the body to learn a lot about.
Write an "All About The..." book for Grade 1-2 students.
Be sure to illustrate your book.

➥

Contact the police to find out how many automobile accidents
there have been in your area this year.
What were the causes of the accidents?
How many fatalities? How many serious injuries?
How might they have been prevented?

➥

Collect newspaper articles about health, nutrition, personal problems.
Write some articles or health advice column letters.

➥

Ask the school nurse to note down for you the various problems
that s/he may encounter in the course of a school day.
(You could provide a form to fill in as each patient visits, so that there will be less to do.)

➥

Check with the emergency ward at a hospital to find out what the most common types of injury are.

➥

Draw a close-up or a cross-section of one of the following:
skin, muscle, bone, eye, ear, heart, lung, etc.

➥

You are a nerve cell in a stubbed toe.
Outline the full process you go through in protecting the toe
and reporting to command central—the brain.

Speech / Drama

Take the role of a part of the body.
Have a discussion between several parts as to why they are the most important.

➥

One person is the heart, a second is the lungs.
The two work together to show how pulse and breathing are related.

➥

Be a reporter.
Interview various organs, muscles or senses to find out how they feel activities such as:
jogging, eating chocolate cake, a detention, being scared.

➥

A foot and a hand discuss their complaints and worries.
A left hand and a right hand compare their problems.

➥

You are a radio reporter.
Describe the blood moving through the body as though it were a Grand Prix race.
Explain each of the main events of each "lap".
Be dramatic!

LOOKING AT BUBBLES

ACTIVITIES FOR THE STUDY OF MATTER (4-7)

UNIT OBJECTIVES

Skills

Observation

* Examine color in bubbles.
* Examine the shapes that bubbles take when in contact.
* Make and examine bubble designs.

Inferring

* Suggest inferences to fit the observation of bubble shapes that are made both with and without apparatus.

Communication

* Draw pictures of multiple bubble patterns.
Describe observations.
* Chart the effects of different bubble mixtures.

Attitudes

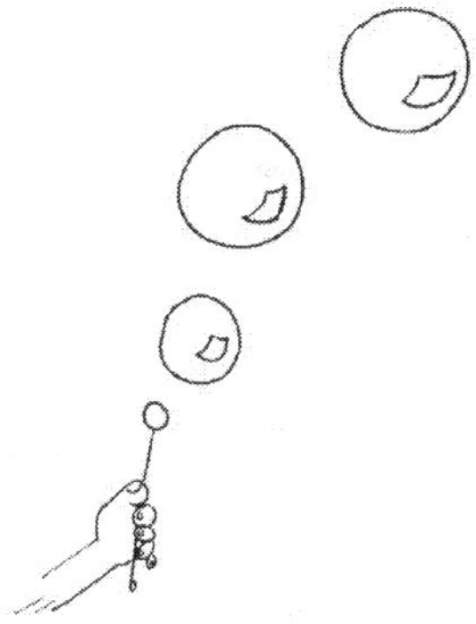

Curiosity

* Examine many ways of making bubbles.

* Study different kinds of bubbles: soap, gum, pop, aerated aquariums, etc.

Concepts

Change

* Bubbles change their shape when in contact with other bubbles or other objects in order to minimize surface area.

Interaction

* Bubbles are made when air is trapped inside something.

* Bubbles can be useful, (e.g., builder's level) or harmful (e.g., the bends).

UNIT ACTIVITIES

Different Detergents

Make a bubble mixture using 1/2 cup (500 mL) liquid dishwater detergent, 4-1/2 cups (4.5 L) water, and 4 tablespoons glycerin (60 mL). Be sure to measure everything very carefully.

Next, make the same quantity of mixture using different detergents.

Does the detergent make any difference to the size or durability of the bubbles?

Make a different set of mixtures using your best detergent and differing quantities of water or glycerine.

Chart your results.

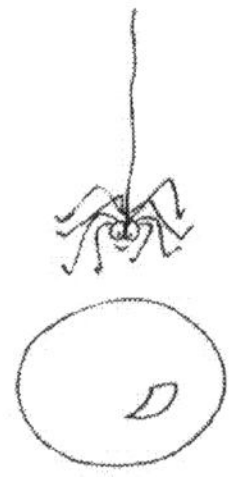

Build Bigger Bubbles

Use blowers of different dimensions made from a variety of materials. (E.g., straws, funnels, wire loops, etc.)

Draw a graph to see if there is a relationship between the diameter of a bubble and the diameter of the blower.

Make Rainbow Bubbles

Bubbles can break up sunlight to make rainbows.

Sit outside in the sunlight and blow some bubbles.

Note each colour as you see it:
RED, ORANGE, YELLOW, GREEN,
BRIGHT BLUE, DARK BLUE, PURPLE.

Keep It Up!

Blow a small bubble into the air and see how long you can keep it up.

See if there is any relationship between the size of a bubble and the length of time you can keep it up.

Follow a small bubble closely in the air for one minute and see where it winds up.

Would another bubble be the same?
Why or why not?

Where do small seeds wind up? Why?

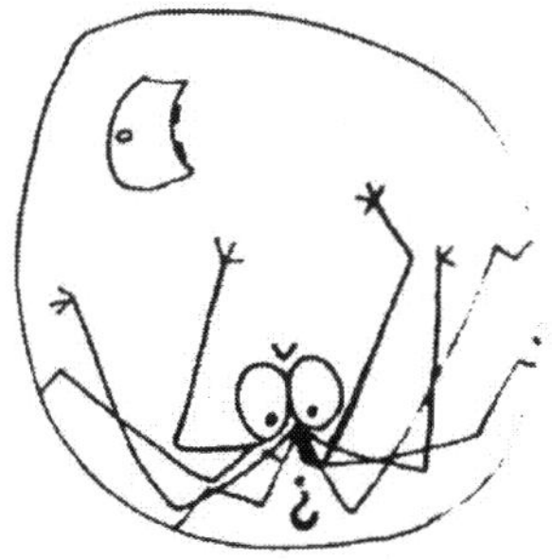

Builders' Bubbles

Fill a screw-top bottle with water and put on the lid.

Make a mark anywhere on the side with a felt pen.

Try to "catch" the mark in the air bubble in the jar.

Make another mark on the side where you will be sure to catch the bubble.

Use your bubble-bottle to find a horizontal surface.

Examine a builder's spirit level.

Bubble Prisms

Make cubes and pyramids with
toothpicks and plasticene.

Attach a thread to each prism shape
and examine what happens after you dip it
into the bubble mixture.

Is the design the same every time
you dip it into the mixture?

Try the experiment with rectangular prisms,
spheres, and triangular prisms.

Other Bubbles

Examine the cause and effect of "the bends" in divers.

Examine how bubbles are formed in soft drinks.

Examine different kinds of bubbles:
bubble gum, balloons,
foam, bubbles in whipped cream,
air bubbles in cake and bread,
bubble wrap for breakables.

Examine how air pockets
are used for insulation.

Make a bubble in water.
How did you do it?
What is inside the bubble that you made?
What happens to the bubble when
it reaches the surface?

SOUND and VIBRATIONS

ACTIVITIES FOR THE STUDY OF ENERGY (4-7)

UNIT OBJECTIVES

Skills

Measuring

* Study differences in pitch and loudness in musical instruments.

* Examine the extent of vibration of different objects when struck.

Classifying

* Discuss the categories of musical instruments.

* Examine objects which reflect or absorb sound.

Space / Time Relationships

* Examine the movement of a vibrating object.

* Examine the sequence of a musical scale.

Attitudes

Persistence

* Examine many aspects of each type of sound.

Inventiveness

* Devise original vibration experiments.

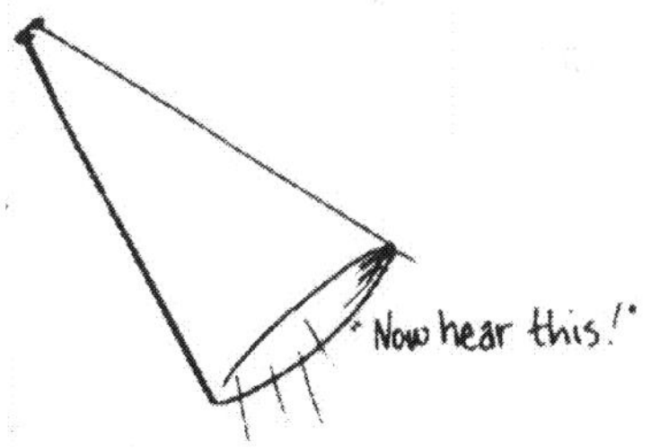

Concepts

Interdependence

* Sound is produced by vibrating objects.

* The pitch of a sound is affected by the characteristics (length, type of material, etc.) of the vibrating object.

Variety / Commonality

* Some substances absorb/reflect sound more readily than others.

UNIT ACTIVITIES

Study Many Different Sounds

Record "mystery" sounds and discuss how they were produced.

Listen to sounds in the environment to identify what is making the sound and how it was produced.

Produce sounds with different objects and determine WHAT is moving to produce sounds, and HOW the movement produces the sound.

The "Bullroarer"

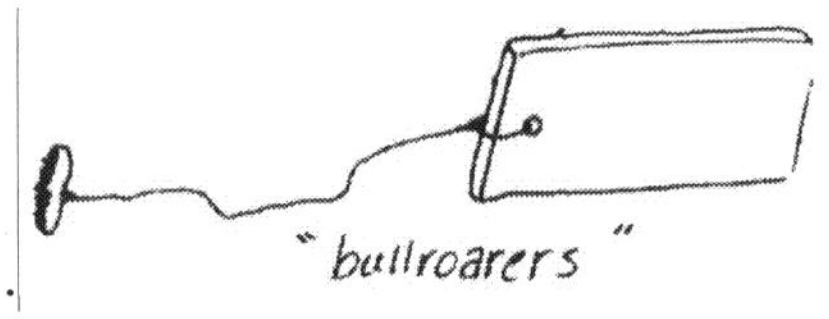

Make a "bullroarer" to investigate how the movement may alter the sound produced (swing quickly, slowly).

Change the width of the bullroarer (thick, thin). Predict how the sounds will differ and test the prediction. Explain why the sounds are different in each case.

Study Pitch

Observe stringed instruments (violin, guitar, ukelele) and discuss how the pitch produced by each string can be changed.

Investigate how the pitch of percussion objects can be controlled.

Make a drum from a burst balloon or a piece of plastic or rubber stretched across a jar, tin, or ice-cream container. Place pebbles or rice grains on your drum. Find ways to alter the pitch.

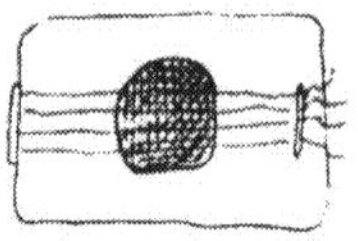

Listen To Bottles

Listen to the differences in the sounds produced when identical bottles with different amounts of water in them are tapped with a pencil. Suspend the bottles with string to accentuate the sound differences.

Which bottle produces the highest / lowest sound?
What is vibrating, the water, the bottle, or the air in the bottle?

Investigate how the sound differs when the same amount of water is poured into a larger / smaller bottle.

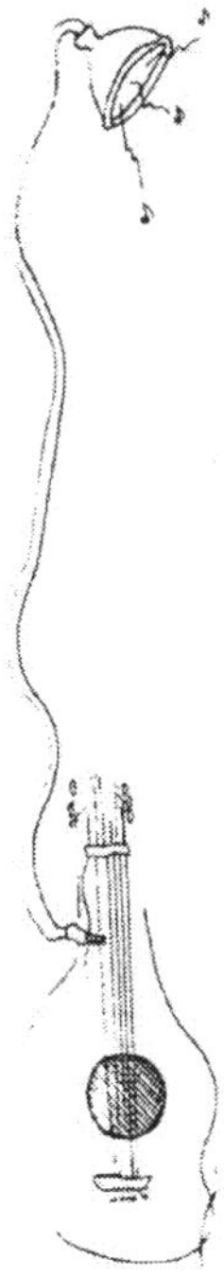

Make Straw Whistles

Use straws to make whistles.
Investigate how pitch can be changed.
Examine straws of different lengths.
Cut finger holes.
Fit the straw into a larger straw to make a slide whistle.

Study Frequency and Pitch

Investigate the relationship between frequency and pitch by fastening a cardboard to the spokes of a wheel and recording what happens as the wheel turns at different speeds.

Study Loudness

Investigate ways of making sounds easier to hear.
Increase the amount of material that vibrates.
Increase the amount of sound collected (control the distance and the volume).
Change the material that the sound travels through.

Study Echoes

Study how echoes are made in a large empty enclosed space such as a hall, stairwell, etc.

What kinds of surfaces make good reflectors?
What things can be used to absorb sound?

Research

- Find out about the keen sense of hearing needed for survival by many animals.
- Find out about noise pollution—around the school, your home, the community.
- Find out about devices used to measure sound.
- Find out how echoes are used in navigation by bats, dolphins etc.
- Find out how the ocean depths are calculated by echo-sounders.
- Observe and research to find out about materials used to insulate wall linings and ceilings in schools, libraries, hospitals, music rooms, etc.
- Find out how a seismograph works.
- Find out about the decible measurement of different noise levels.

NOTES FOR THE TEACHER

VIBRATION: When an object vibrates, it makes the surrounding air vibrate and produce sound. This vibration of the air has two effects:

- As the vibrating object moves outward, it COMPRESSES the surrounding air (compression).
- As the vibrating object moves inward, the air EXPANDS as it spreads out to fill the space formerly occupied by the object (rarefaction).

SOUND WAVES: The compressions and rarefactions of the air near the source of the sound also compress and expand the surrounding air, thus producing SOUND WAVES. The AMPLITUDE and FREQUENCY of a sound wave help to determine the LOUDNESS and PITCH of a sound respectively.

PITCH: If the frequency of vibration is increased, the pitch becomes higher.
The PITCH of a stringed instrument can be altered by changing:

- the TENSION of the strings
- the LENGTH of the strings
- the THICKNESS of the string

The FREQUENCY (PITCH) of sound is measured in hertz units, i.e., cycles per second.

AIR BOTTLES: The shorter the air column in the bottle, the more rapid the vibration and the higher the sound!

LOUDNESS: The LOUDNESS of a sound is the APPARENT strength of the sensation received by the eardrum and sent to the brain. The same AMPLITUDE of sound may produce different degrees of loudness for different people. LOUDNESS depends on:

- The distance of the source of the sound
- The amplitude of the vibrations
- The density of the medium through which the sound travels
- The area of the vibrating object

AMPLITUDE: Sound AMPLITUDE is measured in DECIBLES.

THE LIVING EARTH

ACTIVITIES FOR THE STUDY OF THE EARTH (4-7)

UNIT OBJECTIVES

Skills

Prediction

* Predict the outcome of experiments based on prior knowledge.

Controlling Variables

* Grow seeds under controlled conditions.

Interpreting Data

* Examine weather maps in the daily papers.

Attitudes

Caring

* Show a respect for living things and their environments.

Critical Thinking

* Ensure that experiments are fairly set-up and conducted.

Concepts

Conservation

* Examine ways in which parts of the earth's surface change form.

Continuity

* Examine the overlapping of biomes and habitats.

UNIT ACTIVITIES
Think About The Physical Earth

Examine the Isostasy Theory

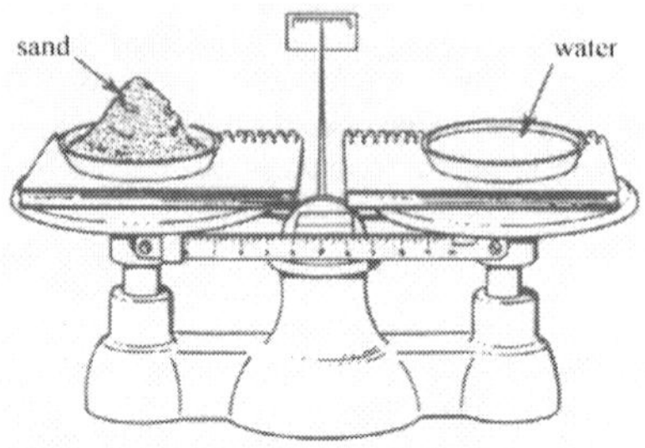

Set up a balance scale as shown in the diagram.
The notebooks should be of the same size and thickness, and almost touching.
These represent the earth's crust.

The pie tins should be centered on the notebooks.
Water in one pie tin represents the oceans. Sand in the other pie tin represents a mountain.

To show that mountain rock is worn down and carried down to the sea,
take a spoonful of sand from the "mountain" and add it to the "ocean".

Observe the motion of the "mountain" and the "ocean".

Examine the Contraction Theory of Earth Movement

Let a hot baked apple cool. Examine the wrinkles and folds.
Compare these to mountains and valleys.

Examine the Convection Theory of Earth Movement

Fill a Pyrex pot with water and add shredded blotting paper.
Stir, and allow the blotting paper to settle to the bottom.
Place it on a hot plate so that only one side touches the element.
Heat the pot and examine the path of the convection current.

Consider how convection currents may affect the earth's crust.

Examine the Theory of Plate Tectonics

Make two oblong blocks of plasticene.
Allow them to cool and harden.
Push the blocks toward each other and note the results.
Look for cracking (earthquakes), folding, and sliding.

Examine Biospheres: Study Environments

Carefully Examine Your House

Look at corners near ceilings, basement floors, and walls,
exposed water pipes and drains, dark attic corners, etc.
Predict what organisms or traces of organisms you might find.
Note any clues (include the location) that show that organisms
other than people and pets live in the house.
Try and find out what organisms there are.

Examine garages, tool sheds, and other places where people do not live.
Make the same kind of list and compare the results.
Make inferences as to why there are differences.

Living Things and Temperature

Place a piece of paper towel into each of three small jars.
Soak the paper towel, and add several bean seeds.
Wrap the jars in foil to keep out all light.

Put one jar in a freezer, one in a refrigerator,
and set one on a table where it can be kept at room temperature.
Label each according to its location. (We are controlling the variables!)

After three days, remove the coverings and examine the growth.
Cover the jars again, and place all three on the table.
Examine again after another three days. Describe and explain any differences.

Examine Plants and Soil

Prepare two shallow boxes by cutting out one end and using a screen to block the opening.
Fill the bottom of each box with soil, and prop at 30 degrees.
Prop them in such a way that the runoff can be collected.
Stick several leaves and twigs into the soil of one box and leave the soil in the other box bare.
With a sprinkler, sprinkle each box with ½ cup (125 mL) of water.
Examine what happens to the soil in each. Measure the amount of water collected from each.

Set up the two boxes again, pressing the soil down firmly.
With a stick, make furrows that run lengthwise in one box, and crosswise in the other.
Sprinkle 1 cup (250 mL) of water onto the high end of each box.
Examine what happens to the soil in each. Measure the amount of water collected from each.
What could this experiment teach a farmer?

Examine Specific Environments

Examine the history of the North American prairies.
Find out about the soil and growth patterns of Amazon jungles.
What effect does burning to clear farm areas have on the soil? Are the farms productive?
Predict what effect tropical rainstorms have on the soil.

Think About Climate

Weather Conditions

How accurate are the daily weather forcasts?
What data do meterologists use to make weather forecasts?
What are the advantages of weather forecasts to fishermen, farmers, pilots?
How do weather forecasts affect our plans and clothing?
How does the weather affect our mood?

Heat and the Weather

Investigate how much heat is absorbed or released by air, water and soil.
Place the substance to be examined over a flame or in the sun.
Measure changes in the temperature at one minute intervals.
Allow the substances to cool. Continue recording temperature change.

Examine the results to determine how quickly heat is absorbed,
how quickly it is released, and how long it is retained.

Examine the temperature of air 10 cm above concrete, sand, water, lawn, metal roofing, etc.
Which substances absorb or release heat most readily?

Experiment With Heated Air

Sprinkle chalk dust above a flame.
(A flashlight beam in a darkened room will aid the observations.)

Hang a paper spiral above a flame.

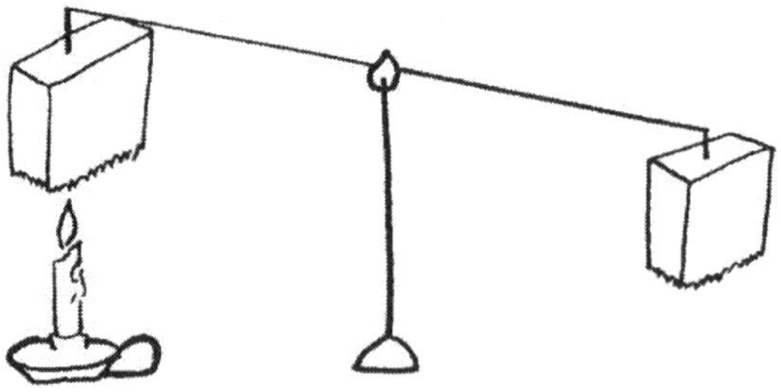

Balance two paper bags upside down, and heat the air under one.

*Hot air does not rise on its own, but is forced upward
by cooler, denser air that settles in and displaces it.*

Climate

Examine differences in the surface of the earth,
and compare the differences in climate in light of the experiments you have done.
Why are there differences in temperature?
Why are there differences in air movement?
What differences are there between coastal and inland climates?
What effects do mountains have on climate?

EXPLORING THE MOON

ACTIVITIES FOR THE STUDY OF SPACE (4-7)

UNIT OBJECTIVES

Skills

Space/Time

* Consider the changes that have taken place in the last 25 years
in our understanding of the moon.

Designing Experiments and Controlling Variables

* Ensure that all experiments are done in a scientific manner,
with accurate measurements,
care that the variables are examined one at a time,
and carefully recorded results from which conclusions can be drawn.

* Plan and design an experiment to study some aspect of space exploration.

Attitudes

Curiosity

* Examine many different ways of learning about the moon.

Objectivity

* Understand the separation between factual knowledge of the moon and speculative fiction.

Concepts

Interdependence

* The moon and the earth affect each other.

Unit Activities
Observe The Moon

Make observations over a period of several days when the Moon is visible in the day time.
Keep a record of its shape, its position in the sky, and the time of the observation.

Visit a local observatory to view the Moon.

List and discuss the needs of life on Earth.
Can the Moon support life?

Find old (prior to 1950) science fiction stories and old science books about the Moon.
How exact were the authors?

Write riddles or jingles using information about space or the Moon.
Write stories and poems about the Moon.

The Moon's Surface

(On the surface of the Moon there are mountains, craters of varying sizes and dry plains, called maria or seas.

Demonstrate the way in which lunar craters may have been formed.

BY VOLCANIC ACTION

Punch a hole in the end of a cardboard box
and push a section of plastic hose through the hole into the center.
Fill the box with loose sandy soil about 7.5 centimeters deep.
Blow very hard on the end of the hose and a crater will form.

BY METEORITES

Drop stones of varying sizes from varying heights onto a soft mud surface.
Observe the depth and size of the "craters" formed.
Toss stones of varying sizes so they will strike the surface from an angle.
How do these craters differ from the craters formed by simply dropping the stones?
What happens if dust, sand, or other material is used instead of mud?
Discuss theories of lunar crater formation.

Make a "moonscape" by placing small objects, strings, etc., on a flat piece of cardboard.
Cover with aluminum foil.
Press foil down firmly around all objects and over the sides of the cardboard.
Paint with black tempera paint to which a small amount of poly-vinyl glue has been added.

Make a three-dimensional map of the Moon's surface.
Draw or paint moonscapes.

Lunar Probes

Notes

Lunar probes were used to collect lunar data before men were sent to the Moon. The series of nine Ranger probes took close-up photographs of the Moon before crashing into the surface. Seven Surveryors soft-landed on the Moon to take pictures and analyze the composition of the surface. Five Lunar Orbiters were placed in low orbit around the Moon to take sharp close-up photographs of extensive areas, to provide information about the Moon's size, shape, and gravitational field, and to determine radiation levels and micrometeorite frequency.

What Facts About the Moon Were Discovered by Space Probes?

Find copies of lunar photographs taken by satellites.
Find examples of several types of craters: primary, secondary, ghost, overlapping.
Compare photographs taken by Ranger probes to those taken by Lunar Orbiters.
Discuss and interpret the details of the Moon's surface.

Preparation for Landing.

What information would have been helpful to have before man landed on the Moon?
How could this information be obtained by the Surveyor?

(Examples are surface hardness, physical and chemical features, seismic activity of the Moon, micrometeorite activity near the surface of the Moon, surface bearing strength, temperatures.)

Demonstrate the Way in Which a Lunar Surface Probe Might Operate.

Place a glass or plastic tube about 30 centimeters long
and about 2.5 cm in diameter upright on the surface to be tested.
Drop a marble or steel ball through the tube onto the surface
and measure the height of the bounce.
Test several different kinds of surfaces and compare.

The ratio between the rebound distance and the height of the drop can be used as an index of surface hardness.

Examine Force.

Push a soda straw through cotton, marshmallows,
rice, coarse gravel, dirt, small rocks, a white potato, etc.
How much force was needed each time?

Repeat using a plastic strip, a wooden strip, and a metal bar instead of the straw.
Compare this to methods used by instruments on Surveyor
to test the Moon's surface composition.

Use a "Strain Gauge" to Test Surface Hardness.

Use an inverted postal weight scale to push a piece of copper tubing or pipe into different kinds of soils.
Record the weight registered on the scale as the amount of force needed to penetrate the soil.
Test several kinds of soil and compare.

Measure Shadow Lengths at Different Times of the Day.

What is the ratio of the height of an object to its shadow?
How does the time of day affect the shadow length?

Make a large model of the Moon's surface.
Place the model in the sunlight and take several pictures from above at different heights.

Photographs should be taken at different times of the day to show that shadow angles and lengths vary with the angle of the sunlight.

Study the photographs and compare them to the model.

The Revolution of the Moon

Look at the moon's orbit.

How long does it take for the moon to complete one orbit around the earth?
What shape is the orbital path?

Show the approximate relationship between the Earth's motion and that of the Moon.

Fold a long sheet of heavy wrapping paper lengthwise down the middle.
Unfold it and place it flat on the floor, so that the fold mark shows.
One child, representing the Earth, walks slowly down the fold line.
Another child, representing the Moon, orbits the Earth, marking his or her path on the paper with a felt-tipped pen.
This must be done carefully, so that the "moon" doesn't try to backtrack.

(fm = full moon)

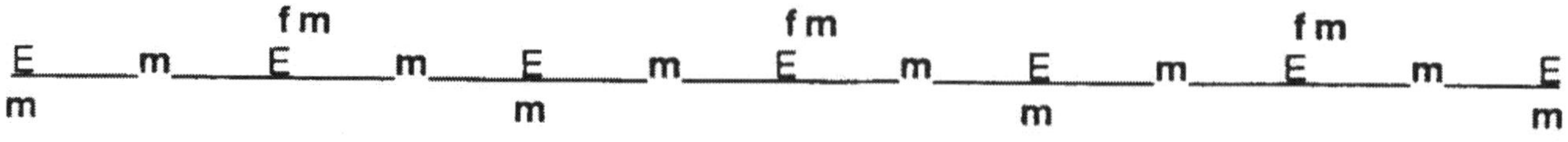

FORCE, FRICTION AND MACHINES

ACTIVITIES FOR THE STUDY OF TECHNOLOGY (4-7)

UNIT OBJECTIVES

Skills

Measurement

* Measure accurately enough to make comparisons.

* Select appropriate units and instruments when measuring.

Communicating

* Learn the rules of brainstorming.

* Explain experimental results carefully and logically.

Controlling Variables

* Be aware of what is being tested.

* Record changes and the effect of the changes.

Attitudes

Curiosity

* Try to find things out for yourself,
by exploring until you feel that you understand them.

Inventiveness

* Suggest alternate ways of solving a problem or extending an activity.
* Work without requiring constant teacher direction.

Honesty

* Report all observations in a truthful manner, even if they seem incorrect.

Concepts

Cause and Effect

* Changes in the amount of friction measured
are due to changes in the testing conditions.

Interaction

* Friction is both helpful and harmful to us.

Unit Activities
Brute Force

Examine Energy

Discuss the word "energy".
What gives living things energy?
When do living things use energy?
Describe how you feel when you're "full of energy".
Describe how you feel when you haven't any energy.

List foods that give energy.

Think of objects that are moved by people physically.
What parts of your body do you use to move things?
How can you tell when you have used energy?
List many things that happen to your body.

List five examples of things that are hard to do.
List five examples of things that are easy to do.
List five things that you do every day that you think are work.
(Not schoolwork!)
Would you call these activities work:
swimming, skipping, running, playing the piano?

Whenever anything is moved, some FORCE has been exerted to cause the movement.
Even when something is stopped, force had to be exerted.
In fact, every time any WORK is done, force is used.
You may consider some of these activities as only play,
but a scientist would call them all types of work, because force has to be exerted even to move a finger.

Have a Tug of War With a Friend

What happens if you pull harder than your friend?
What happens if your friend pulls harder?
Draw diagrams to show both situations.
Draw an arrow above the pictures to show in which direction
each of you is pulling.
Make the arrow longer in the direction that you are moving.
What happens if you both pull with equal force?
Draw a diagram to show the direction of your pull.
Did you draw both arrows of equal length?

If an object is not moving, it is not because no forces are acting on it.
If equal and opposite forces act on an object, it will remain without movement.

Friction: It's a Drag!

Rub your hands together briskly.
Change the method of rubbing.
Increase the pressure between your hands.
Increase the speed of rubbing.
Decrease the amount of surface touching.
(E.g., use only the finger tips, heel of the hand, palm, etc.)
Hold one hand still.
What effect did each of the changes have on your ability to rub your hands?

Rub your hand over a piece of sandpaper or rough wood.
Now rub it over a piece of glass or mirror.
Over which surface did your hand slide more easily?

Try to pull a heavy object along a cement path or along the ground.
Now try pulling it along a polished linoleum floor or other smooth surface.
On which surface was it easier to pull?
Why do you think this was so?

Using Spring Balances to Test Friction

Use a spring balance to measure how much effort is required
to drag a heavy book across a desk top.
Repeat the activity to find the force required to pull the same book
across other surfaces. (glass, carpet, sandpaper)

Use a hand lens or microscope to carefully examine the surfaces being tested.
What are the differences between the surfaces which required the
least / greatest effort to move the book?

Use a spring balance to measure how much effort is required
to drag a rectangular block of wood across the table.
Turn the block on its edge so that there is less surface area touching the test surface
and compare the effort required to drag it.

On rough surfaces, your hand seems to "drag" instead of slipping easily. This drag is called FRICTION. Even the glass or mirror over which your hand slipped relatively easily is not completely smooth. If you were to look at it under a microscope, you would see tiny lumps and bumps. Nothing in the world is perfectly smooth.

Pushing and Lifting

Try pushing a very heavy book across the floor.
Then put three long pencils under it and try pushing the book again.
Was it easier to push the first or the second time? Why?

Rollers and Wheels

What would primitive man have used for rollers as you used the pencils?
How would it make moving things easier if the logs are shaped
so that only the two ends touch the ground?

Find out how this evolved into the axle and wheels of today.

Is friction a good or bad thing to have when we want to move something,
such as a bicycle or skateboard? Why?

Do spokes on a wheel reduce friction between the ground and the wheel?
Be sure to explain your answer.

Why does an inflated basketball roll further than a slightly flat one?

Levers

Wheels were not the only energy-saving idea that early man had.
When something very heavy was to be moved often LEVER was used.
A lever consists of a rigid bar or rod such as a crowbar, resting on a fixed point.

You can find out more about the principle of balance and levers by using a seesaw.
A bar used as a lever is really a seesaw.
The box is the weight on one end, or the RESISTANCE.
The weight on the other end is actually the EFFORT used by the person
who pushes down on the lever.
The point that the seesaw balances on is the FULCRUM.

Is it easier to walk carrying two heavy shopping bags in the same hand,
or with one in each hand? Why?
Is it easier to walk with one heavy bag or with two? Why?

Name four other ways you have seen levers used.

Something Extra!

Show how to make fire by rubbing two sticks together.

For centuries men have, from time to time, claimed to have made
perpetual motion machines. Such machines are supposed to run forever
after being started. Do you think this is possible? Why or why not?

SECTION 9 - SCIENCE CERTIFICATES

SKILLS

Observer of Scientific Phenomena
(Observation, Inference, & Prediction)

Analyst of Experimental Processes
(Measuring, Classifying, & Interpreting Data)

Designer of Experimental Projects
(Creating Operational Definitions & Formulating Hypotheses)

Coordinator of Communications
(Verbal, Written, & Graphic Communications)

Manager of Space-Time Relationships

CONTENT

Animal Awareness
Plant Proficiency
Energy Enigma
Master of Matter
Earth Expert
Space Scholar
Technology Trustee
Health & Physiology Practitioner

ATTITUDE

Scientific Attitude

Having shown ability in the areas of

Observation, Inference, & Prediction

____

is hereby declared to be a

Skilled Observer of Scientific Phenomena

In confirmation whereof,
I have hereto affixed my signature.

Dated this ___ day of _____, 20__

____

authorized signature

Having shown ability in the areas of

Measuring, Classifying, & Interpreting Data

____

is hereby declared to be a

Skilled Analyst
of
Experimental Processes

In confirmation whereof,
I have hereto affixed my signature.

Dated this ___ day of _____, 20__

____

authorized signature

Having shown ability in the areas of

Creating Operational Definitions & Formulating Hypotheses

____

is hereby declared to be a

Skilled Designer
of
Experimental Projects

In confirmation whereof,
I have hereto affixed my signature.

Dated this ___ day of _____, 20__

____

authorized signature

Having shown ability in

Verbal, Written, & Graphic Communications

____

is henceforth deemed to be a

Co-Ordinator of Communications

In confirmation whereof,
I have hereto affixed my signature.

Dated this ___ day of _____, 20__

____

authorized signature

Having shown an understanding of

Size, Shape, and Position

and Changes in these over Time

____

is henceforth declared to be an

Skilled Manager
of
Space & Time Relationships

In confirmation whereof,
I have hereto affixed my signature.

Dated this ___ day of ______, 20__

____

authorized signature

Having shown

Care & Concern
for all Kinds of Animals

____

is henceforth proclaimed to have a superior level of

Animal Awareness

In confirmation whereof,
I have hereto affixed my signature.

Dated this ___ day of _____, 20__

____

authorized signature

For having shown an understanding of

Plant Growth and Care

____

is henceforth proclaimed to have a superior level of

Plant Proficiency

In confirmation whereof,
I have hereto affixed my signature.

Dated this ___ day of _____, 20__

____

authorized signature

Having shown an understanding of

Sound and Vibration

is hereinafter entitled to be called an

Energy Enigma

In confirmation whereof,
I have hereto affixed my signature.

Dated this ___ day of _____, 20__

authorized signature

Having shown an understanding of

Heat, Light & Sound

____

is hereinafter entitled to be called an

Energy Egghead

In confirmation whereof,
I have hereto affixed my signature.

Dated this ___ day of _____, 20__

____

authorized signature

Having shown an understanding of

Potential and Kinetic Energy

____

is hereinafter entitled to be called an

Master of Matter

In confirmation whereof,
I have hereto affixed my signature.

Dated this ___ day of _____, 20__

____

authorized signature

Having demonstrated an understanding of

Natural Processes of the Earth

____

is henceforth declared to be an

Earth Expert

In confirmation whereof,
I have hereto affixed my signature.

Dated this ___ day of _____, 20__

____

authorized signature

Having demonstrated an understanding of

The Earth, the Solar System, and the Vast Reaches of the Universe Beyond

____

is henceforth declared to be an

Space Scholar

In confirmation whereof,
I have hereto affixed my signature.

Dated this ___ day of _____, 20__

____

authorized signature

Having shown understanding and concern in the areas of

Physical, Social, & Emotional Health

____

is entitled to receive the designation of

Health & Physiology Practitioner

In confirmation whereof,
I have hereto affixed my signature.

Dated this ___ day of _____, 20__

____

authorized signature

This is to certify that

____

has been shown to be a

Technology Trustee

capable of understanding
the Modern World
and the many machines and systems
that help to make it run smoothly.

In confirmation whereof,
I have hereto affixed my signature.

Dated this ___ day of _____, 20__

____

authorized signature

____

is hereby declared to be

The Proud Possessor

of a

Scientific Attitude

Having shown due measure of the following:

Curiosity Caring Critical Thinking

Persistence Inventiveness

In confirmation whereof,
I have hereto affixed my signature.

Dated this ___ day of _____, 20__

____

authorized signature

Science works...
... works!

* see inside for details

Made in United States
Troutdale, OR
11/03/2026

56844643R00166